THE EXPRESSION OF THE EMOTIONS IN MAN AND ANIMALS

The Expression of the Emotions in MAN AND ANIMALS

CHARLES DARWIN

WITH A PREFACE
BY KONRAD LORENZ

THE UNIVERSITY OF CHICAGO PRESS
CHICAGO & LONDON

Reprinted from the Authorized Edition of D. Appleton and Company, New York and London

ISBN: 0-226-13655-8 (clothbound); 0-226-13656-6 (paperbound)
THE UNIVERSITY OF CHICAGO PRESS, CHICAGO 60637
The University of Chicago Press, Ltd., London
 Published 1965. Fourth Impression 1970
Printed in the United States of America

CONTENTS

ILLUSTRATIONS

Figures

Plates

PREFACE

BY KONRAD LORENZ

A very great biologist and highly respected teacher of mine, Jacob von Uexküll, once said rather pessimistically that today's truth was, after all, nothing but the error of tomorrow. Thereupon, another great biologist who is also one of my most highly revered teachers, Otto Koehler, answered: "No, the truth of today is the special case of tomorrow!" Unquestionably, this second statement contains a very much deeper truth than the first. In science, and particularly in biology, the discoverer of a new explanatory principle is more than apt to overrate the range of its applicability. When Jacques Loeb discovered the principle of tropisms, he more than half believed—and hoped—that all animal and human behavior could be explained on the basis of interacting tropisms. When I. P. Pavlov discovered the conditioned response, he more or less believed the same thing of his explanatory principle. The writings of Sigmund Freud are full of analogous generalizations. One may indulgently regard this little weakness as the well-merited prerogative of genius, because the great man's pupils, though lesser discoverers, are apt to be better at verification than their inspired teacher and can be relied upon to clip the wings of his genius when it threatens to soar too high. It is only when the pupils degenerate into disciples who unquestioningly accept the far-sweeping statements of their master that danger arises, and that a newly born epistemophagous (knowledge-devouring) monster, another "ism," rears its ugly head.

However, the greatest of all discoverers in the field of biology did not commit the error just discussed: when Charles Darwin discovered natural selection, the explanatory principle that was destined to change our outlook on man and the world more than any other before it, he decidedly did not overestimate the number of phenomena that could be explained on its basis. If anything, he erred on the side of understatement. For this reason, I very much resent the term "Darwinism" as an unjust aspersion that accuses the great man of a sin which he, of all people, would have abhorred.

Modern biologists are much more "Darwinistic" than Darwin, and with reason. We are more insistent in our quest for a definite selection pressure whenever one of nature's more elaborate constructions arouses our curiosity and our demand for a causal explanation. Since the days of Darwin, repeated success has given us great confidence that, whenever one of evolution's more intriguing products provides us with a puzzle, a diligent and circumspect search for specific selection pressure will provide us with a solution. I would go so far as to assert that any, even the most strikingly unbelievable, forms of structure and behavior can be understood, at least in principle, as the outcome of the selection pressure exerted by their particular survival function. We are always ready to ask the question "What for?" which, with us, does not imply a profession of mystical teleology. When we ask: "What does a cat have crooked, retractile claws for?" and answer: "To catch mice with!" we just state, in a short way, that catching mice was the primary function whose enormous survival value has bred cats with that particular formation of claws, and that it has done so by the same process of selection by which a human breeder produces strains of chickens with an enormous egg production or Pekinese with infinitesimal noses.

Like all really great scientific discoverers, Darwin possessed an almost uncanny ability to reason on the basis of hypotheses which were not only provisional and vague but also subconscious. He deduced correct consequences from facts more suspected than known, and verified both the theory and the facts by the obvious truth of the conclusions thus reached. In other words, a man like Darwin knows much more than he thinks he knows, and it is not surprising that the consequences of his knowledge reach far and in different directions. Many different branches of biological research have been inspired by him, and each is claiming him, with equal right, as its particular originator and pioneer. What is surprising is the extent to which further research, based on Darwin's hypotheses and pursuing them in every conceivable direction, has invariably proved him right on every essential point.

The branch of behavior study commonly called ethology, which can be defined succinctly as the biology of behavior, has a special right to claim Charles Darwin as its patron saint. It is more immediately dependent on the selectionistic approach than any other biological science I could name, and it has done its fair share in verifying Darwin's theories. Furthermore, in his book, *The Expression of the Emotions in Man and Animals*, Charles Darwin has foreseen in a truly visionary manner the main problems which confront ethologists to this day and has mapped out a strategy of research which they still use. Reading between the lines, one can see that Darwin was fully aware of a fact which, though simple in itself, is so fundamental to biological behavior study that its rediscovery by Charles Otis Whitman and Oskar Heinroth is rightly considered the starting point of ethology.

This fact, which is still ignored by many psychologists, is quite simply that behavior patterns are just as conservatively and reliably characters of species as are the forms of bones, teeth, or any other bodily structures. Similarities in inherited behavior unite the members of a species, of a genus, and of even the largest taxonomic units in exactly the same way in which bodily characters do so. The conservative persistence of behavior patterns, even after they have outlived, in the evolution of a species, their original function, is exactly the same as that of organs; in other words, they can become "vestigial" or "rudimentary," just as the latter can. Or, on losing one function, they may develop another, as the first gill slit became an ear opening when our ancestors changed from aquatic to terrestrial life. Darwin shows in the most convincing manner that analogous processes have taken place in the evolution of motor patterns, as for instance, in the case of "snarling," in which an expression movement with a purely communicative function has developed out of the motor pattern of actual biting which, as a means of aggression, has practically disappeared in the human species.

That behavior patterns have an evolution exactly like that of organs is a fact which entails the recognition of another: that they also have the same sort of heredity.

In other words, the adaptation of the behavior patterns of an organism to its environment is achieved in exactly the same manner as that of its organs, that is to say on the basis of information which the species has gained in the course of its evolution by the age-old method of mutation and selection. This is true not only for relatively rigid patterns of form or behavior, but also for the complicated mechanisms of adaptive modification, among which are those generally subsumed under the conception of learning.

In my opinion, it is in the field of behavior study that the undeniable truths contained in *The Expression of the Emotions in Man and Animals* develop their most far-reaching consequences, theoretically, practically, and even politically. From this conviction I derive the inner justification for writing this Preface. I do not doubt, however, that many other biologists, working in different fields, might claim the same with equal right. I believe that even today we do not quite realize how much Charles Darwin knew.

INTRODUCTION.

MANY works have been written on Expression, but a greater number on Physiognomy,—that is, on the recognition of character through the study of the permanent form of the features. With this latter subject I am not here concerned. The older treatises,[1] which I have consulted, have been of little or no service to me. The famous 'Conférences'[2] of the painter Le Brun, published in 1667, is the best known ancient work, and contains some good remarks. Another somewhat old essay, namely, the 'Discours,' delivered 1774–1782, by the well-known Dutch anatomist Camper,[3] can hardly be considered as having made any marked advance in the subject. The following works, on the contrary, deserve the fullest consideration.

Sir Charles Bell, so illustrious for his discoveries in physiology, published in 1806 the first edition, and in

[1] J. Parsons, in his paper in the Appendix to the 'Philosophical Transactions' for 1746, p. 41, gives a list of forty-one old authors who have written on Expression.

[2] 'Conférences sur l'expression des différents Caractères des Passions.' Paris, 4to, 1667. I always quote from the republication of the 'Conférences' in the edition of Lavater, by Moreau, which appeared in 1820, as given in vol. ix. p. 257.

[3] 'Discours par Pierre Camper sur le moyen de représenter les diverses passions,' &c. 1792.

1844 the third edition of his 'Anatomy and Philosophy of Expression.'[4] He may with justice be said, not only to have laid the foundations of the subject as a branch of science, but to have built up a noble structure. His work is in every way deeply interesting; it includes graphic descriptions of the various emotions, and is admirably illustrated. It is generally admitted that his service consists chiefly in having shown the intimate relation which exists between the movements of expression and those of respiration. One of the most important points, small as it may at first appear, is that the muscles round the eyes are involuntarily contracted during violent expiratory efforts, in order to protect these delicate organs from the pressure of the blood. This fact, which has been fully investigated for me with the greatest kindness by Professors Donders of Utrecht, throws, as we shall hereafter see, a flood of light on several of the most important expressions of the human countenance. The merits of Sir C. Bell's work have been undervalued or quite ignored by several foreign writers, but have been fully admitted by some, for instance by M. Lemoine,[5] who with great justice says: —"Le livre de Ch. Bell devrait être médité par quiconque essaye de faire parler le visage de l'homme, par les philosophes aussi bien que par les artistes, car, sous une apparence plus légère et sous le prétexte de l'esthétique, c'est un des plus beaux monuments de la science des rapports du physique et du moral."

From reasons which will presently be assigned, Sir

---

[4] I always quote from the third edition, 1844, which was published after the death of Sir C. Bell, and contains his latest corrections. The first edition of 1806 is much inferior in merit, and does not include some of his more important views.

[5] 'De la Physionomie et de la Parole,' par Albert Lemoine, 1865, p. 101.

C. Bell did not attempt to follow out his views as far as they might have been carried. He does not try to explain why different muscles are brought into action under different emotions; why, for instance, the inner ends of the eyebrows are raised, and the corners of the mouth depressed, by a person suffering from grief or anxiety.

In 1807 M. Moreau edited an edition of Lavater on Physiognomy,[6] in which he incorporated several of his own essays, containing excellent descriptions of the movements of the facial muscles, together with many valuable remarks. He throws, however, very little light on the philosophy of the subject. For instance, M. Moreau, in speaking of the act of frowning, that is, of the contraction of the muscle called by French writers the *sourcilier* (*corrugator supercilii*), remarks with truth:—"Cette action des sourciliers est un des symp-

[6] 'L'Art de connaître les Hommes,' &c., par G. Lavater. The earliest edition of this work, referred to in the preface to the edition of 1820 in ten volumes, as containing the observations of M. Moreau, is said to have been published in 1807; and I have no doubt that this is correct, because the 'Notice sur Lavater' at the commencement of volume i. is dated April 13, 1806. In some bibliographical works, however, the date of 1805--1809 is given, but it seems impossible that 1805 can be correct. Dr. Duchenne remarks ('Mécanisme de la Physionomie Humaine,' 8vo edit. 1862, p. 5, and 'Archives Générales de Médecine,' Jan. et Fév. 1862) that M. Moreau "*a composé pour son ouvrage un article important*," &c., in the year 1805; and I find in volume i. of the edition of 1820 passages bearing the dates of December 12, 1805, and another January 5, 1806, besides that of April 13, 1806, above referred to. In consequence of some of these passages having thus been *composed* in 1805, Dr. Duchenne assigns to M. Moreau the priority over Sir C. Bell, whose work, as we have seen, was published in 1806. This is a very unusual manner of determining the priority of scientific works; but such questions are of extremely little importance in comparison with their relative merits. The passages above quoted from M. Moreau and from Le Brun are taken in this and all other cases from the edition of 1820 of Lavater, tom. iv. p. 228, and tom. ix. p. 279.

tômes les plus tranchés de l'expression des affections pénibles ou concentrées." He then adds that these muscles, from their attachment and position, are fitted "à resserrer, à concentrer les principaux traits de la *face*, comme il convient dans toutes ces passions vraiment oppressives ou profondes, dans ces affections dont le sentiment semble porter l'organisation à revenir sur elle-même, à se contracter et à *s'amoindrir*, comme pour offrir moins de prise et de surface à des impressions redoutables ou importunes." He who thinks that remarks of this kind throw any light on the meaning or origin of the different expressions, takes a very different view of the subject to what I do.

In the above passage there is but a slight, if any, advance in the philosophy of the subject, beyond that reached by the painter Le Brun, who, in 1667, in describing the expression of fright, says:—"Le sourcil qui est abaissé d'un côté et élevé de l'autre, fait voir que la partie élevée semble le vouloir joindre au cerveau pour le garantir du mal que l'âme aperçoit, et le côté qui est abaissé et qui paraît enflé, nous fait trouver dans cet état par les esprits qui viennent du cerveau en abondance, comme pour couvrir l'ame et la défendre du mal qu'elle craint; la bouche fort ouverte fait voir le saisissement du cœur, par le sang qui se retire vers lui, ce qui l'oblige, voulant respirer, à faire un effort qui est cause que la bouche s'ouvre extrêmement, et qui, lorsqu'il passe par les organes de la voix, forme un son qui n'est point articulé; que si les muscles et les veines paraissent enflés, ce n'est que par les esprits que le cerveau envoie en ces parties-là." I have thought the foregoing sentences worth quoting, as specimens of the surprising nonsense which has been written on the subject.

'The Physiology or Mechanism of Blushing,' by Dr.

Burgess, appeared in 1839, and to this work I shall frequently refer in my thirteenth Chapter.

In 1862 Dr. Duchenne published two editions, in folio and octavo, of his 'Mécanisme de la Physionomie Humaine,' in which he analyses by means of electricity, and illustrates by magnificent photographs, the movements of the facial muscles. He has generously permitted me to copy as many of his photographs as I desired. His works have been spoken lightly of, or quite passed over, by some of his countrymen. It is possible that Dr. Duchenne may have exaggerated the importance of the contraction of single muscles in giving expression; for, owing to the intimate manner in which the muscles are connected, as may be seen in Henle's anatomical drawings [7]—the best I believe ever published —it is difficult to believe in their separate action. Nevertheless, it is manifest that Dr. Duchenne clearly apprehended this and other sources of error, and as it is known that he was eminently successful in elucidating the physiology of the muscles of the hand by the aid of electricity, it is probable that he is generally in the right about the muscles of the face. In my opinion, Dr. Duchenne has greatly advanced the subject by his treatment of it. No one has more carefully studied the contraction of each separate muscle, and the consequent furrows produced on the skin. He has also, and this is a very important service, shown which muscles are least under the separate control of the will. He enters very little into theoretical considerations, and seldom attempts to explain why certain muscles and not others contract under the influence of certain emotions.

A distinguished French anatomist, Pierre Gratiolet,

[7] 'Handbuch der Systematischen Anatomie des Menschen.' Band I. Dritte Abtheilung, 1858.

gave a course of lectures on Expression at the Sorbonne, and his notes were published (1865) after his death, under the title of 'De la Physionomie et des Mouvements d'Expression.' This is a very interesting work, full of valuable observations. His theory is rather complex, and, as far as it can be given in a single sentence (p. 65), is as follows:—"Il résulte, de tous les faits que j'ai rappelés, que les sens, l'imagination et la pensée elle-même, si élevée, si abstraite qu'on la suppose, ne peuvent s'exercer sans éveiller un sentiment corrélatif, et que ce sentiment se traduit directement, sympathiquement, symboliquement ou métaphoriquement, dans toutes les sphères des organs extérieurs, qui la racontent tous, suivant leur mode d'action propre, comme si chacun d'eux avait été directement affecté."

Gratiolet appears to overlook inherited habit, and even to some extent habit in the individual; and therefore he fails, as it seems to me, to give the right explanation, or any explanation at all, of many gestures and expressions. As an illustration of what he calls symbolic movements, I will quote his remarks (p. 37), taken from M. Chevreul, on a man playing at billiards. "Si une bille dévie légèrement de la direction que le joueur prétend lui imprimer, ne l'avez-vous pas vu cent fois la pousser du regard, de la tête et même des épaules, comme si ces mouvements, purement symboliques, pouvaient rectifier son trajet? Des mouvements non moins significatifs se produisent quand la bille manque d'une impulsion suffisante. Et chez les joueurs novices, ils sont quelquefois accusés au point d'éveiller le sourire sur les lèvres des spectateurs." Such movements, as it appears to me, may be attributed simply to habit. As often as a man has wished to move an object to one side, he has always pushed it to that side; when forwards, he has pushed it forwards; and if he has wished to arrest it, he has pulled backwards. Therefore, when a man sees his ball travelling in a wrong direction, and he intensely wishes it to go in another direction, he cannot avoid, from long habit, unconsciously performing movements which in other cases he has found effectual.

As an instance of sympathetic movements Gratiolet gives (p. 212) the following case:—"un jeune chien à oreilles droites, auquel son maître présente de loin quelque viande appétissante, fixe avec ardeur ses yeux sur cet objet dont il suit tous les mouvements, et pendant que les yeux regardent, les deux oreilles se portent en avant comme si cet objet pouvait être entendu." Here, instead of speaking of sympathy between the ears and eyes, it appears to me more simple to believe, that as dogs during many generations have, whilst intently looking at any object, pricked their ears in order to perceive any sound; and conversely have looked intently in the direction of a sound to which they may have listened, the movements of these organs have become firmly associated together through long-continued habit.

Dr. Piderit published in 1859 an essay on Expression, which I have not seen, but in which, as he states, he forestalled Gratiolet in many of his views. In 1867 he published his 'Wissenschaftliches System der Mimik und Physiognomik.' It is hardly possible to give in a few sentences a fair notion of his views; perhaps the two following sentences will tell as much as can be briefly told: "the muscular movements of expression are in part related to imaginary objects, and in part to imaginary sensorial impressions. In this proposition lies the key to the comprehension of all expressive muscular movements." (s. 25.) Again, "Expressive movements manifest themselves chiefly in the numerous and mobile muscles of the face, partly because the nerves by which they are set into motion originate in the most immediate vicinity of the mind-organ, but partly also because these muscles serve to support the organs of sense." (s. 26.) If Dr. Piderit had studied Sir C. Bell's work, he would probably not have said (s. 101) that violent laughter causes a frown from partaking of the nature of pain; or that with infants (s. 103) the tears irritate the eyes, and thus excite the contraction of the surrounding muscles. Many good remarks are scattered throughout this volume, to which I shall hereafter refer.

Short discussions on Expression may be found in various works, which need not here be particularised. Mr. Bain, however, in two of his works has treated the subject at some length. He says,[8] "I look upon the expression so-called as part and parcel of the feeling. I believe it to be a general law of the mind that, along with the fact of inward feeling or consciousness, there is a diffusive action or excitement over the bodily members." In another place he adds, "A very considerable number of the facts may be brought under the following principle: namely, that states of pleasure are connected with an increase, and states of pain with an abatement, of some, or all, of the vital functions." But the above law of the diffusive action of feelings seems too general to throw much light on special expressions.

Mr. Herbert Spencer, in treating of the Feelings in his 'Principles of Psychology' (1855), makes the following remarks:—"Fear, when strong, expresses itself in cries, in efforts to hide or escape, in palpitations and tremblings; and these are just the manifestations that

[8] 'The Senses and the Intellect,' 2nd edit. 1864, pp. 96 and 288. The preface to the first edition of this work is dated June, 1855. See also the 2nd edition of Mr. Bain's work on the 'Emotions and Will.'

would accompany an actual experience of the evil feared. The destructive passions are shown in a general tension of the muscular system, in gnashing of the teeth and protrusion of the claws, in dilated eyes and nostrils, in growls; and these are weaker forms of the actions that accompany the killing of prey." Here we have, as I believe, the true theory of a large number of expressions; but the chief interest and difficulty of the subject lies in following out the wonderfully complex results. I infer that some one (but who he is I have not been able to ascertain) formerly advanced a nearly similar view, for Sir C. Bell says,[9] "It has been maintained that what are called the external signs of passion, are only the concomitants of those voluntary movements which the structure renders necessary." Mr. Spencer has also published [10] a valuable essay on the physiology of Laughter, in which he insists on "the general law that feeling passing a certain pitch, habitually vents itself in bodily action;" and that "an overflow of nerve-force undirected by any motive, will manifestly take first the most habitual routes; and if these do not suffice, will next overflow into the less habitual ones." This law I believe to be of the highest importance in throwing light on our subject.[11]

---

[9] 'The Anatomy of Expression,' 3rd edit. p. 121.

[10] 'Essays, Scientific, Political, and Speculative,' Second Series, 1863, p. 111. There is a discussion on Laughter in the First Series of Essays, which discussion seems to me of very inferior value.

[11] Since the publication of the essay just referred to, Mr. Spencer has written another, on "Morals and Moral Sentiments," in the 'Fortnightly Review,' April 1, 1871, p. 426. He has, also, now published his final conclusions in vol. ii. of the second edit. of the 'Principles of Psychology,' 1872, p. 539. I may state, in order that I may not be accused of trespassing on Mr. Spencer's domain, that I announced in my 'Descent of Man,' that I had then written a part of the present volume: my first MS. notes on the subject of expression bear the date of the year 1838.

All the authors who have written on Expression, with the exception of Mr. Spencer—the great expounder of the principle of Evolution—appear to have been firmly convinced that species, man of course included, came into existence in their present condition. Sir C. Bell, being thus convinced, maintains that many of our facial muscles are "purely instrumental in expression;" or are "a special provision" for this sole object.[12] But the simple fact that the anthropoid apes possess the same facial muscles as we do,[13] renders it very improbable that these muscles in our case serve exclusively for expression; for no one, I presume, would be inclined to admit that monkeys have been endowed with special muscles solely for exhibiting their hideous grimaces. Distinct uses, independently of expression, can indeed be assigned with much probability for almost all the facial muscles.

Sir C. Bell evidently wished to draw as broad a distinction as possible between man and the lower animals; and he consequently asserts that with "the lower creatures there is no expression but what may be referred, more or less plainly, to their acts of volition or necessary instincts." He further maintains that their faces "seem chiefly capable of expressing rage and fear."[14] But man himself cannot express love and humility by external signs, so plainly as does a dog, when with drooping ears, hanging lips, flexuous body, and wagging tail, he meets his beloved master. Nor can these movements

[12] 'Anatomy of Expression,' 3rd edit. pp. 98, 121, 131.

[13] Professor Owen expressly states (Proc. Zoolog. Soc. 1830, p. 28) that this is the case with respect to the Orang, and specifies all the more important muscles which are well known to serve with man for the expression of his feelings. See, also, a description of several of the facial muscles in the Chimpanzee, by Prof. Macalister, in 'Annals and Magazine of Natural History,' vol. vii. May, 1871, p. 342.

[14] 'Anatomy of Expression,' pp. 121, 138.

in the dog be explained by acts of volition or necessary instincts, any more than the beaming eyes and smiling cheeks of a man when he meets an old friend. If Sir C. Bell had been questioned about the expression of affection in the dog, he would no doubt have answered that this animal had been created with special instincts, adapting him for association with man, and that all further enquiry on the subject was superfluous.

Although Gratiolet emphatically denies [15] that any muscle has been developed solely for the sake of expression, he seems never to have reflected on the principle of evolution. He apparently looks at each species as a separate creation. So it is with the other writers on Expression. For instance, Dr. Duchenne, after speaking of the movements of the limbs, refers to those which give expression to the face, and remarks: [16] "Le créateur n'a donc pas eu à se préoccuper ici des besoins de la mécanique; il a pu, selon sa sagesse, ou—que l'on me pardonne cette manière de parler—par une divine fantaisie, mettre en action tel ou tel muscle, un seul ou plusieurs muscles à la fois, lorsqu'il a voulu que les signes caractéristiques des passions, même les plus fugaces, fussent écrits passagèrement sur la face de l'homme. Ce langage de la physionomie une fois créé, il lui a suffi, pour le rendre universel et immuable, de donner à tout être humain la faculté instinctive d'exprimer toujours ses sentiments par la contraction des mêmes muscles."

Many writers consider the whole subject of Expression as inexplicable. Thus the illustrious physiologist Müller, says,[17] "The completely different expression of

---

[15] 'De la Physionomie,' pp. 12, 73.

[16] 'Mécanisme de la Physionomie Humaine,' 8vo edit. p. 31.

[17] 'Elements of Physiology,' English translation, vol. ii. p. 934.

the features in different passions shows that, according to the kind of feeling excited, entirely different groups of the fibres of the facial nerve are acted on. Of the cause of this we are quite ignorant."

No doubt as long as man and all other animals are viewed as independent creations, an effectual stop is put to our natural desire to investigate as far as possible the causes of Expression. By this doctrine, anything and everything can be equally well explained; and it has proved as pernicious with respect to Expression as to every other branch of natural history. With mankind some expressions, such as the bristling of the hair under the influence of extreme terror, or the uncovering of the teeth under that of furious rage, can hardly be understood, except on the belief that man once existed in a much lower and animal-like condition. The community of certain expressions in distinct though allied species, as in the movements of the same facial muscles during laughter by man and by various monkeys, is rendered somewhat more intelligible, if we believe in their descent from a common progenitor. He who admits on general grounds that the structure and habits of all animals have been gradually evolved, will look at the whole subject of Expression in a new and interesting light.

The study of Expression is difficult, owing to the movements being often extremely slight, and of a fleeting nature. A difference may be clearly perceived, and yet it may be impossible, at least I have found it so, to state in what the difference consists. When we witness any deep emotion, our sympathy is so strongly excited, that close observation is forgotten or rendered almost impossible; of which fact I have had many curious proofs. Our imagination is another and still more serious source of error; for if from the nature of the circumstances we expect to see any expression, we readily imagine its presence. Notwithstanding Dr. Duchenne's great experience, he for a long time fancied, as he states, that several muscles contracted under certain emotions, whereas he ultimately convinced himself that the movement was confined to a single muscle.

In order to acquire as good a foundation as possible, and to ascertain, independently of common opinion, how far particular movements of the features and gestures are really expressive of certain states of the mind, I have found the following means the most serviceable. In the first place, to observe infants; for they exhibit many emotions, as Sir C. Bell remarks, "with extraordinary force;" whereas, in after life, some of our expressions "cease to have the pure and simple source from which they spring in infancy."[18]

In the second place, it occurred to me that the insane ought to be studied, as they are liable to the strongest passions, and give uncontrolled vent to them. I had, myself, no opportunity of doing this, so I applied to Dr. Maudsley and received from him an introduction to Dr. J. Crichton Browne, who has charge of an immense asylum near Wakefield, and who, as I found, had already attended to the subject. This excellent observer has with unwearied kindness sent me copious notes and descriptions, with valuable suggestions on many points; and I can hardly over-estimate the value of his assistance. I owe also, to the kindness of Mr. Patrick Nicol, of the Sussex Lunatic Asylum, interesting statements on two or three points.

Thirdly Dr. Duchenne galvanized, as we have already seen, certain muscles in the face of an old man, whose skin was little sensitive, and thus produced various ex-

---

[18] 'Anatomy of Expression,' 3rd edit. p. 198.

pressions which were photographed on a large scale. It fortunately occurred to me to show several of the best plates, without a word of explanation, to above twenty educated persons of various ages and both sexes, asking them, in each case, by what emotion or feeling the old man was supposed to be agitated; and I recorded their answers in the words which they used. Several of the expressions were instantly recognised by almost everyone, though described in not exactly the same terms; and these may, I think, be relied on as truthful, and will hereafter be specified. On the other hand, the most widely different judgments were pronounced in regard to some of them. This exhibition was of use in another way, by convincing me how easily we may be misguided by our imagination; for when I first looked through Dr. Duchenne's photographs, reading at the same time the text, and thus learning what was intended, I was struck with admiration at the truthfulness of all, with only a few exceptions. Nevertheless, if I had examined them without any explanation, no doubt I should have been as much perplexed, in some cases, as other persons have been.

Fourthly, I had hoped to derive much aid from the great masters in painting and sculpture, who are such close observers. Accordingly, I have looked at photographs and engravings of many well-known works; but, with a few exceptions, have not thus profited. The reason no doubt is, that in works of art, beauty is the chief object; and strongly contracted facial muscles destroy beauty.[19] The story of the composition is generally told with wonderful force and truth by skilfully given accessories.

Fifthly, it seemed to me highly important to ascer-

---

[19] See remarks to this effect in Lessing's 'Laocoon,' translated by W. Ross, 1836, p. 19.

tain whether the same expressions and gestures prevail, as has often been asserted without much evidence, with all the races of mankind, especially with those who have associated but little with Europeans. Whenever the same movements of the features or body express the same emotions in several distinct races of man, we may infer with much probability, that such expressions are true ones,—that is, are innate or instinctive. Conventional expressions or gestures, acquired by the individual during early life, would probably have differed in the different races, in the same manner as do their languages. Accordingly I circulated, early in the year 1867, the following printed queries with a request, which has been fully responded to, that actual observations, and not memory, might be trusted. These queries were written after a considerable interval of time, during which my attention had been otherwise directed, and I can now see that they might have been greatly improved. To some of the later copies, I appended, in manuscript, a few additional remarks:—

(1.) Is astonishment expressed by the eyes and mouth being opened wide, and by the eyebrows being raised?

(2.) Does shame excite a blush when the colour of the skin allows it to be visible? and especially how low down the body does the blush extend?

(3.) When a man is indignant or defiant does he frown, hold his body and head erect, square his shoulders and clench his fists?

(4.) When considering deeply on any subject, or trying to understand any puzzle, does he frown, or wrinkle the skin beneath the lower eyelids?

(5.) When in low spirits, are the corners of the mouth depressed, and the inner corner of the eyebrows raised by that muscle which the French call the "Grief muscle"? The eyebrow in this state becomes slightly oblique, with a little swelling at the inner end; and the forehead is transversely wrinkled in the middle part, but not across the whole breadth, as when the eyebrows are raised in surprise.

(6.) When in good spirits do the eyes sparkle, with the skin a little wrinkled round and under them, and with the mouth a little drawn back at the corners?

(7.) When a man sneers or snarls at another, is the corner of the upper lip over the canine or eye tooth raised on the side facing the man whom he addresses?

(8.) Can a dogged or obstinate expression be recognized, which is chiefly shown by the mouth being firmly closed, a lowering brow and a slight frown?

(9.) Is contempt expressed by a slight protrusion of the lips and by turning up the nose, and with a slight expiration?

(10.) Is disgust shown by the lower lip being turned down, the upper lip slightly raised, with a sudden expiration, something like incipient vomiting, or like something spit out of the mouth?

(11.) Is extreme fear expressed in the same general manner as with Europeans?

(12.) Is laughter ever carried to such an extreme as to bring tears into the eyes?

(13.) When a man wishes to show that he cannot prevent something being done, or cannot himself do something, does he shrug his shoulders, turn inwards his elbows, extend outwards his hands and open the palms; with the eyebrows raised?

(14.) Do the children when sulky, pout or greatly protrude the lips?

(15.) Can guilty, or sly, or jealous expressions be recognized? though I know not how these can be defined.

(16.) Is the head nodded vertically in affirmation, and shaken laterally in negation?

Observations on natives who have had little communication with Europeans would be of course the most valuable, though those made on any natives would be of much interest to me. General remarks on expression are of comparatively little value; and memory is so deceptive that I earnestly beg it may not be trusted. A definite description of the countenance under any emotion or frame of mind, with a statement of the circumstances under which it occurred, would possess much value.

To these queries I have received thirty-six answers from different observers, several of them missionaries or protectors of the aborigines, to all of whom I am deeply indebted for the great trouble which they have taken, and for the valuable aid thus received. I will specify their names, &c., towards the close of this chapter, so as not to interrupt my present remarks. The answers relate to several of the most distinct and savage races of man. In many instances, the circumstances have been recorded under which each expression was observed, and the expression itself described. In such cases, much confidence may be placed in the answers. When the answers have been simply yes or no, I have always received them with caution. It follows, from the information thus acquired, that the same state of mind is expressed throughout the world with remarkable uniformity; and this fact is in itself interesting as evidence of the close similarity in bodily structure and mental disposition of all the races of mankind.

Sixthly, and lastly, I have attended, as closely as I could, to the expression of the several passions in some of the commoner animals; and this I believe to be of paramount importance, not of course for deciding how far in man certain expressions are characteristic of certain states of mind, but as affording the safest basis for generalisation on the causes, or origin, of the various movements of Expression. In observing animals, we are not so likely to be biassed by our imagination; and we may feel safe that their expressions are not conventional.

From the reasons above assigned, namely, the fleeting nature of some expressions (the changes in the features being often extremely slight); our sympathy being easily aroused when we behold any strong emotion, and our attention thus distracted; our imagination deceiving us, from knowing in a vague manner what to expect, though certainly few of us know what the exact changes in the countenance are; and lastly, even our long familiarity with the subject,—from all these causes combined, the observation of Expression is by no means easy, as many persons, whom I have asked to observe certain points, have soon discovered. Hence it is difficult to determine, with certainty, what are the movements of the features and of the body, which commonly characterize certain states of the mind. Nevertheless, some of the doubts and difficulties have, as I hope, been cleared away by the observation of infants,—of the insane,—of the different races of man,—of works of art,—and lastly, of the facial muscles under the action of galvanism, as effected by Dr. Duchenne.

But there remains the much greater difficulty of understanding the cause or origin of the several expressions, and of judging whether any theoretical explanation is trustworthy. Besides, judging as well as we can by our reason, without the aid of any rules, which of two or more explanations is the most satisfactory, or are quite unsatisfactory, I see only one way of testing our conclusions. This is to observe whether the same principle by which one expression can, as it appears, be explained, is applicable in other allied cases; and especially, whether the same general principles can be applied with satisfactory results, both to man and the lower animals. This latter method, I am inclined to think, is the most serviceable of all. The difficulty of judging of the truth of any theoretical explanation, and of testing it by some distinct line of investigation, is the great drawback to that interest which the study seems well fitted to excite.

Finally, with respect to my own observations, I may state that they were commenced in the year 1838; and from that time to the present day, I have occasionally attended to the subject. At the above date, I was already inclined to believe in the principle of evolution, or of the derivation of species from other and lower forms. Consequently, when I read Sir C. Bell's great work, his view, that man had been created with certain muscles specially adapted for the expression of his feelings, struck me as unsatisfactory. It seemed probable that the habit of expressing our feelings by certain movements, though now rendered innate, had been in some manner gradually acquired. But to discover how such habits had been acquired was perplexing in no small degree. The whole subject had to be viewed under a new aspect, and each expression demanded a rational explanation. This belief led me to attempt the present work, however imperfectly it may have been executed.

I will now give the names of the gentlemen to whom, as I have said, I am deeply indebted for information in regard to the expressions exhibited by various races of man, and I will specify some of the circumstances under which the observations were in each case made. Owing to the great kindness and powerful influence of Mr. Wilson, of Hayes Place, Kent, I have received from Australia no less than thirteen sets of answers to my queries. This has been particularly fortunate, as the Australian aborigines rank amongst the most distinct of all the races of man. It will be seen that the observations have been chiefly made in the south, in the outlying parts of the colony of Victoria; but some excellent answers have been received from the north.

Mr. Dyson Lacy has given me in detail some valuable observations, made several hundred miles in the interior of Queensland. To Mr. R. Brough Smyth, of Melbourne, I am much indebted for observations made by himself, and for sending me several of the following letters, namely:—From the Rev. Mr. Hagenauer, of

Lake Wellington, a missionary in Gippsland, Victoria, who has had much experience with the natives. From Mr. Samuel Wilson, a landowner, residing at Langerenong, Wimmera, Victoria. From the Rev. George Taplin, superintendent of the native Industrial Settlement at Port Macleay. From Mr. Archibald G. Lang, of Coranderik, Victoria, a teacher at a school where aborigines, old and young, are collected from all parts of the colony. From Mr. H. B. Lane, of Belfast, Victoria, a police magistrate and warden, whose observations, as I am assured, are highly trustworthy. From Mr. Templeton Bunnett, of Echuca, whose station is on the borders of the colony of Victoria, and who has thus been able to observe many aborigines who have had little intercourse with white men. He compared his observations with those made by two other gentlemen long resident in the neighbourhood. Also from Mr. J. Bulmer, a missionary in a remote part of Gippsland, Victoria.

I am also indebted to the distinguished botanist, Dr. Ferdinand Müller, of Victoria, for some observations made by himself, and for sending me others made by Mrs. Green, as well as for some of the foregoing letters.

In regard to the Maoris of New Zealand, the Rev. J. W. Stack has answered only a few of my queries; but the answers have been remarkably full, clear, and distinct, with the circumstances recorded under which the observations were made.

The Rajah Brooke has given me some information with respect to the Dyaks of Borneo.

Respecting the Malays, I have been highly successful; for Mr. F. Geach (to whom I was introduced by Mr. Wallace), during his residence as a mining engineer in the interior of Malacca, observed many natives, who had never before associated with white men. He wrote me two long letters with admirable and detailed observations on their expression. He likewise observed the Chinese immigrants in the Malay archipelago.

The well-known naturalist, H. M. Consul, Mr. Swinhoe, also observed for me the Chinese in their native country; and he made inquiries from others whom he could trust.

In India Mr. H. Erskine, whilst residing in his official capacity in the Admednugur District in the Bombay Presidency, attended to the expression of the inhabitants, but found much difficulty in arriving at any safe conclusions, owing to their habitual concealment of all emotions in the presence of Europeans. He also obtained information for me from Mr. West, the Judge in Canara, and he consulted some intelligent native gentlemen on certain points. In Calcutta Mr. J. Scott, curator of the Botanic Gardens, carefully observed the various tribes of men therein employed during a considerable period, and no one has sent me such full and valuable details. The habit of accurate observation, gained by his botanical studies, has been brought to bear on our present subject. For Ceylon I am much indebted to the Rev. S. O. Glenie for answers to some of my queries.

Turning to Africa, I have been unfortunate with respect to the negroes, though Mr. Winwood Reade aided me as far as lay in his power. It would have been comparatively easy to have obtained information in regard to the negro slaves in America; but as they have long associated with white men, such observations would have possessed little value. In the southern parts of the continent Mrs. Barber observed the Kafirs and Fingoes, and sent me many distinct answers. Mr. J. P. Mansel Weale also made some observations on the natives, and procured for me a curious document, namely, the opinion, written in English, of Christian Gaika, brother of the Chief Sandilli, on the expressions of his fellow-countrymen. In the northern regions of Africa Captain Speedy, who long resided with the Abyssinians, answered my queries partly from memory and partly from observations made on the son of King Theodore, who was then under his charge. Professor and Mrs. Asa Gray attended to some points in the expressions of the natives, as observed by them whilst ascending the Nile.

On the great American continent Mr. Bridges, a catechist residing with the Fuegians, answered some few questions about their expression, addressed to him many years ago. In the northern half of the continent Dr. Rothrock attended to the expressions of the wild Atnah and Espyox tribes on the Nasse River, in North-Western America. Mr. Washington Matthews, Assistant-Surgeon in the United States Army, also observed with special care (after having seen my queries, as printed in the 'Smithsonian Report') some of the wildest tribes in the Western parts of the United States, namely, the Tetons, Grosventres, Mandans, and Assinaboines; and his answers have proved of the highest value.

Lastly, besides these special sources of information, I have collected some few facts incidentally given in books of travels.

As I shall often have to refer, more especially in the latter part of this volume, to the muscles of the human face, I have had a diagram (fig. 1) copied and reduced from Sir C. Bell's work, and two others, with more accurate details (figs. 2 and 3), from Henle's well-known 'Handbuch der Systematischen Anatomie des Menschen.' The same letters refer to the same muscles in all three figures, but the names are given of only the more important ones to which I shall have to allude. The facial muscles blend much together, and, as I am informed, hardly appear on a dissected face so distinct as they are here represented. Some writers consider that these muscles consist of nineteen pairs, with one unpaired;[20] but others make the number much larger, amounting even to fifty-five, according to Moreau. They are, as is admitted by everyone who has written on the subject, very variable in structure; and Moreau remarks that they are hardly alike in half-a-dozen subjects.[21] They are also variable in function. Thus the power of uncovering the canine tooth on one side differs much in different persons. The power of raising the wings of the nostrils is also, according to Dr. Piderit,[22] variable in a remarkable degree; and other such cases could be given.

Finally, I must have the pleasure of expressing my obligations to Mr. Rejlander for the trouble which he has taken in photographing for me various expressions and gestures. I am also indebted to Herr Kindermann, of Hamburg, for the loan of some excellent negatives of crying infants; and to Dr. Wallich for a charming one of a smiling girl. I have already expressed my obligations to Dr. Duchenne for generously permitting me to have some of his large photographs copied and reduced. All these photographs have been printed by the Heliotype process, and the accuracy of the copy is thus guaranteed. These plates are referred to by Roman numerals.

I am also greatly indebted to Mr. T. W. Wood for

[20] Mr. Partridge in Todd's 'Cyclopædia of Anatomy and Physiology,' vol. ii. p. 227.

[21] 'La Physionomie,' par G. Lavater, tom. iv. 1820, p. 274. On the number of the facial muscles, see vol. iv. pp. 209--211.

[22] 'Mimik und Physiognomik,' 1867, s. 91.

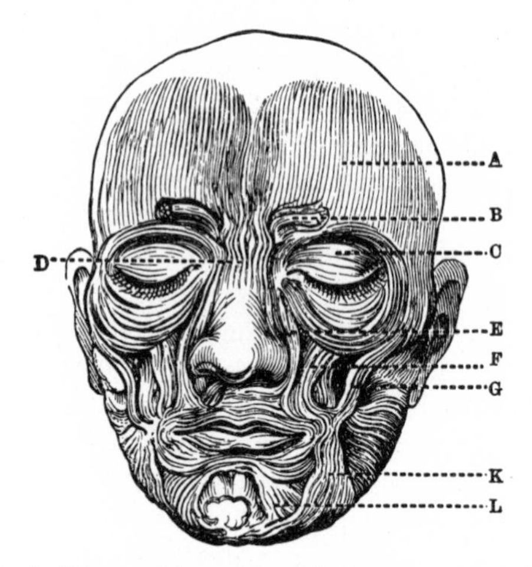

FIG. 1.—Diagram of the muscles of the face, from Sir C. Bell.

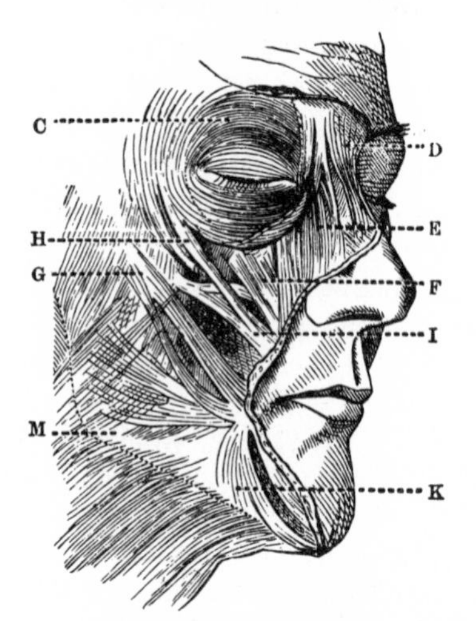

FIG. 2.—Diagram from Henle.

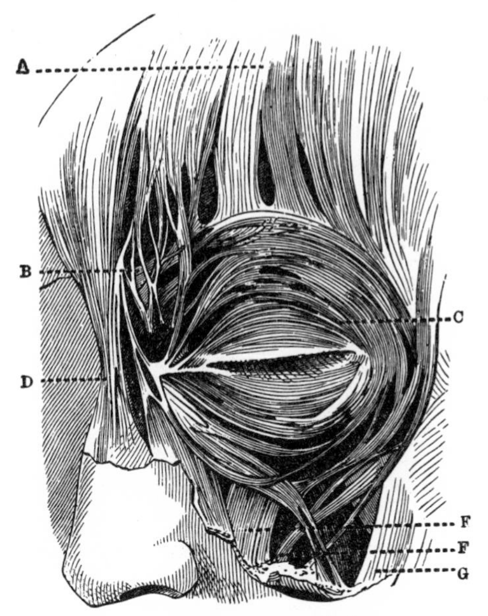

FIG. 3.—Diagram from Henle.

A. Occipito-frontalis, or frontal muscle.
B. Corrugator supercilii, or corrugator muscle.
C. Orbicularis palpebrarum, or orbicular muscles of the eyes.
D. Pyramidalis nasi, or pyramidal muscle of the nose.
E. Levator labii superioris alæque nasi.
F. Levator labii proprius.
G. Zygomatic.
H. Malaris.
I. Little zygomatic.
K. Triangularis oris, or depressor anguli oris.
L. Quadratus menti.
M. Risorius, part of the Platysma myoides.

the extreme pains which he has taken in drawing from life the expressions of various animals. A distinguished artist, Mr. Riviere, has had the kindness to give me two drawings of dogs—one in a hostile and the other in a humble and caressing frame of mind. Mr. A. May has also given me two similar sketches of dogs. Mr. Cooper has taken much care in cutting the blocks. Some of

3

the photographs and drawings, namely, those by Mr. May, and those by Mr. Wolf of the Cynopithecus, were first reproduced by Mr. Cooper on wood by means of photography, and then engraved: by this means almost complete fidelity is ensured.

CHAPTER I.

General Principles of Expression.

The three chief principles stated—The first principle—Serviceable actions become habitual in association with certain states of the mind, and are performed whether or not of service in each particular case—The force of habit—Inheritance—Associated habitual movements in man—Reflex actions—Passage of habits into reflex actions—Associated habitual movements in the lower animals—Concluding remarks.

I WILL begin by giving the three Principles, which appear to me to account for most of the expressions and gestures involuntarily used by man and the lower animals, under the influence of various emotions and sensations.[1] I arrived, however, at these three Principles only at the close of my observations. They will be discussed in the present and two following chapters in a general manner. Facts observed both with man and the lower animals will here be made use of; but the latter facts are preferable, as less likely to deceive us. In the fourth and fifth chapters, I will describe the special expressions of some of the lower animals; and in the succeeding chapters those of man. Every-one will thus be able to judge for himself, how far my

[1] Mr. Herbert Spencer ('Essays,' Second Series, 1863, p. 138) has drawn a clear distinction between emotions and sensations, the latter being "generated in our corporeal framework." He classes as Feelings both emotions and sensations.

three principles throw light on the theory of the subject. It appears to me that so many expressions are thus explained in a fairly satisfactory manner, that probably all will hereafter be found to come under the same or closely analogous heads. I need hardly premise that movements or changes in any part of the body,—as the wagging of a dog's tail, the drawing back of a horse's ears, the shrugging of a man's shoulders, or the dilatation of the capillary vessels of the skin,—may all equally well serve for expression. The three Principles are as follows.

I. *The principle of serviceable associated Habits.*—Certain complex actions are of direct or indirect service under certain states of the mind, in order to relieve or gratify certain sensations, desires, &c.; and whenever the same state of mind is induced, however feebly, there is a tendency through the force of habit and association for the same movements to be performed, though they may not then be of the least use. Some actions ordinarily associated through habit with certain states of the mind may be partially repressed through the will, and in such cases the muscles which are least under the separate control of the will are the most liable still to act, causing movements which we recognize as expressive. In certain other cases the checking of one habitual movement requires other slight movements; and these are likewise expressive.

II. *The principle of Antithesis.*—Certain states of the mind lead to certain habitual actions, which are of service, as under our first principle. Now when a directly opposite state of mind is induced, there is a strong and involuntary tendency to the performance of movements of a directly opposite nature, though these are of no use; and such movements are in some cases highly expressive.

III. *The principle of actions due to the constitution of the Nervous System, independently from the first of the Will, and independently to a certain extent of Habit.*—When the sensorium is strongly excited, nerve-force is generated in excess, and is transmitted in certain definite directions, depending on the connection of the nerve-cells, and partly on habit: or the supply of nerve-force may, as it appears, be interrupted. Effects are thus produced which we recognize as expressive. This third principle may, for the sake of brevity, be called that of the direct action of the nervous system.

With respect to our *first Principle*, it is notorious how powerful is the force of habit. The most complex and difficult movements can in time be performed without the least effort or consciousness. It is not positively known how it comes that habit is so efficient in facilitating complex movements; but physiologists admit[2] "that the conducting power of the nervous fibres increases with the frequency of their excitement." This applies to the nerves of motion and sensation, as well as to those connected with the act of thinking. That some physical change is produced in the nerve-cells or nerves which are habitually used can hardly be doubted, for otherwise it is impossible to understand how the tendency to certain acquired movements is inherited. That they are inherited we see with horses in certain transmitted paces, such as cantering and ambling, which are not natural to them,—in the pointing of young pointers and the setting of young setters—in the peculiar

[2] Müller, 'Elements of Physiology,' Eng. translat. vol. ii. p. 939. See also Mr. H. Spencer's interesting speculations on the same subject, and on the genesis of nerves, in his 'Principles of Biology,' vol. ii. p. 346; and in his 'Principles of Psychology,' 2nd edit. pp. 511--557.

manner of flight of certain breeds of the pigeon, &c. We have analogous cases with mankind in the inheritance of tricks or unusual gestures, to which we shall presently recur. To those who admit the gradual evolution of species, a most striking instance of the perfection with which the most difficult consensual movements can be transmitted, is afforded by the humming-bird Sphinx-moth (*Macroglossa*); for this moth, shortly after its emergence from the cocoon, as shown by the bloom on its unruffled scales, may be seen poised stationary in the air, with its long hair-like proboscis uncurled and inserted into the minute orifices of flowers; and no one, I believe, has ever seen this moth learning to perform its difficult task, which requires such unerring aim.

When there exists an inherited or instinctive tendency to the performance of an action, or an inherited taste for certain kinds of food, some degree of habit in the individual is often or generally requisite. We find this in the paces of the horse, and to a certain extent in the pointing of dogs; although some young dogs point excellently the first time they are taken out, yet they often associate the proper inherited attitude with a wrong odour, and even with eyesight. I have heard it asserted that if a calf be allowed to suck its mother only once, it is much more difficult afterwards to rear it by hand.[3] Caterpillars which have been fed on the leaves of one kind of tree, have been known to perish from hunger rather than to eat the leaves of another tree, although this afforded them their proper food,

[3] A remark to much the same effect was made long ago by Hippocrates and by the illustrious Harvey; for both assert that a young animal forgets in the course of a few days the art of sucking, and cannot without some difficulty again acquire it. I give these assertions on the authority of Dr. Darwin, 'Zoonomia,' 1794, vol. i. p. 140.

under a state of nature;[4] and so it is in many other cases.

The power of Association is admitted by everyone. Mr. Bain remarks, that "actions, sensations and states of feeling, occurring together or in close succession, tend to grow together, or cohere, in such a way that when any one of them is afterwards presented to the mind, the others are apt to be brought up in idea."[5] It is so important for our purpose fully to recognize that actions readily become associated with other actions and with various states of the mind, that I will give a good many instances, in the first place relating to man, and afterwards to the lower animals. Some of the instances are of a very trifling nature, but they are as good for our purpose as more important habits. It is known to everyone how difficult, or even impossible it is, without repeated trials, to move the limbs in certain opposed directions which have never been practised. Analogous cases occur with sensations, as in the common experiment of rolling a marble beneath the tips of two crossed fingers, when it feels exactly like two marbles. Everyone protects himself when falling to the ground by extending his arms, and as Professor Alison has remarked, few can resist acting thus, when voluntarily falling on a soft bed. A man when going out of doors puts on his gloves quite unconsciously; and this may seem an extremely simple operation, but he who has taught a child

[4] See for my authorities, and for various analogous facts, 'The Variation of Animals and Plants under Domestication,' 1868, vol. ii. p. 304.

[5] 'The Senses and the Intellect,' 2nd edit. 1864, p. 332. Prof. Huxley remarks ('Elementary Lessons in Physiology,' 5th edit. 1872, p. 306), "It may be laid down as a rule, that, if any two mental states be called up together, or in succession, with due frequency and vividness, the subsequent production of the one of them will suffice to call up the other, and that whether we desire it or not."

to put on gloves, knows that this is by no means the case.

When our minds are much affected, so are the movements of our bodies; but here another principle besides habit, namely the undireeted overflow of nerve-force, partially comes into play. Norfolk, in speaking of Cardinal Wolsey, says—

> " Some strange commotion
> Is in his brain; he bites his lip and starts;
> Stops on a sudden, looks upon the ground,
> Then, lays his finger on his temple: straight,
> Springs out into fast gait; then, stops again,
> Strikes his breast hard; and anon, he casts
> His eye against the moon: in most strange postures
> We have seen him set himself."—*Hen. VIII.*, act 3, sc. 2.

A vulgar man often scratches his head when perplexed in mind; and I believe that he acts thus from habit, as if he experienced a slightly uncomfortable bodily sensation, namely, the itching of his head, to which he is particularly liable, and which he thus relieves. Another man rubs his eyes when perplexed, or gives a little cough when embarrassed, acting in either case as if he felt a slightly uncomfortable sensation in his eyes or windpipe.[6]

From the continued use of the eyes, these organs are especially liable to be acted on through association under various states of the mind, although there is manifestly nothing to be seen. A man, as Gratiolet remarks, who vehemently rejects a proposition, will almost certainly shut his eyes or turn away his face; but if he accepts the proposition, he will nod his head in affirmation and open his eyes widely. The man acts in this

---

[6] Gratiolet ('De la Physionomie,' p. 324), in his discussion on this subject, gives many analogous instances. See p. 42, on the opening and shutting of the eyes. Engel is quoted (p. 323) on the changed paces of a man, as his thoughts change.

latter case as if he clearly saw the thing, and in the former case as if he did not or would not see it. I have noticed that persons in describing a horrid sight often shut their eyes momentarily and firmly, or shake their heads, as if not to see or to drive away something disagreeable; and I have caught myself, when thinking in the dark of a horrid spectacle, closing my eyes firmly. In looking suddenly at any object, or in looking all around, everyone raises his eyebrows, so that the eyes may be quickly and widely opened; and Duchenne remarks that [7] a person in trying to remember something often raises his eyebrows, as if to see it. A Hindoo gentleman made exactly the same remark to Mr. Erskine in regard to his countrymen. I noticed a young lady earnestly trying to recollect a painter's name, and she first looked to one corner of the ceiling and then to the opposite corner, arching the one eyebrow on that side; although, of course, there was nothing to be seen there.

In most of the foregoing cases, we can understand how the associated movements were acquired through habit; but with some individuals, certain strange gestures or tricks have arisen in association with certain states of the mind, owing to wholly inexplicable causes, and are undoubtedly inherited. I have elsewhere given one instance from my own observation of an extraordinary and complex gesture, associated with pleasurable feelings, which was transmitted from a father to his daughter, as well as some other analogous facts.[8]

---

[7] 'Mécanisme de la Physionomie Humaine,' 1862, p. 17.

[8] 'The Variation of Animals and Plants under Domestication,' vol. ii. p. 6. The inheritance of habitual gestures is so important for us, that I gladly avail myself of Mr. F. Galton's permission to give in his own words the following remarkable case:—" The following account of a habit occurring in individuals of three consecutive generations

Another curious instance of an odd inherited movement, associated with the wish to obtain an object, will be given in the course of this volume.

There are other actions which are commonly performed under certain circumstances, independently of habit, and which seem to be due to imitation or some sort of sympathy. Thus persons cutting anything with

---

is of peculiar interest, because it occurs only during sound sleep, and therefore cannot be due to imitation, but must be altogether natural. The particulars are perfectly trustworthy, for I have enquired fully into them, and speak from abundant and independent evidence. A gentleman of considerable position was found by his wife to have the curious trick, when he lay fast asleep on his back in bed, of raising his right arm slowly in front of his face, up to his forehead, and then dropping it with a jerk, so that the wrist fell heavily on the bridge of his nose. The trick did not occur every night, but occasionally, and was independent of any ascertained cause. Sometimes it was repeated incessantly for an hour or more. The gentleman's nose was prominent, and its bridge often became sore from the blows which it received. At one time an awkward sore was produced, that was long in healing, on account of the recurrence, night after night, of the blows which first caused it. His wife had to remove the button from the wrist of his night-gown as it made severe scratches, and some means were attempted of tying his arm.

"Many years after his death, his son married a lady who had never heard of the family incident. She, however, observed precisely the same peculiarity in her husband; but his nose, from not being particularly prominent, has never as yet suffered from the blows. The trick does not occur when he is half-asleep, as, for example, when dozing in his arm-chair, but the moment he is fast asleep it is apt to begin. It is, as with his father, intermittent; sometimes ceasing for many nights, and sometimes almost incessant during a part of every night. It is performed, as it was by his father, with his right hand.

"One of his children, a girl, has inherited the same trick. She performs it, likewise, with the right hand, but in a slightly modified form; for, after raising the arm, she does not allow the wrist to drop upon the bridge of the nose, but the palm of the half-closed hand falls over and down the nose, striking it rather rapidly. It is also very intermittent with this child, not occurring for periods of some months, but sometimes occurring almost incessantly."

a pair of scissors may be seen to move their jaws simultaneously with the blades of the scissors. Children learning to write often twist about their tongues as their fingers move, in a ridiculous fashion. When a public singer suddenly becomes a little hoarse, many of those present may be heard, as I have been assured by a gentleman on whom I can rely, to clear their throats; but here habit probably comes into play, as we clear our own throats under similar circumstances. I have also been told that at leaping matches, as the performer makes his spring, many of the spectators, generally men and boys, move their feet; but here again habit probably comes into play, for it is very doubtful whether women would thus act.

Reflex actions.—Reflex actions, in the strict sense of the term, are due to the excitement of a peripheral nerve, which transmits its influence to certain nerve-cells, and these in their turn excite certain muscles or glands into action; and all this may take place without any sensation or consciousness on our part, though often thus accompanied. As many reflex actions are highly expressive, the subject must here be noticed at some little length. We shall also see that some of them graduate into, and can hardly be distinguished from actions which have arisen through habit.[9] Coughing and sneezing are familiar instances of reflex actions. With infants the first act of respiration is often a sneeze, although this requires the co-ordinated movement of

[9] Prof. Huxley remarks ('Elementary Physiology,' 5th edit. p. 305) that reflex actions proper to the spinal cord are *natural*; but, by the help of the brain, that is through habit, an infinity of *artificial* reflex actions may be acquired. Virchow admits ('Sammlung wissenschaft. Vorträge,' &c., "Ueber das Rückenmark," 1871, ss. 24, 31) that some reflex actions can hardly be distinguished from instincts; and, of the latter, it may be added, some cannot be distinguished from inherited habits.

numerous muscles. Respiration is partly voluntary, but mainly reflex, and is performed in the most natural and best manner without the interference of the will. A vast number of complex movements are reflex. As good an instance as can be given is the often-quoted one of a decapitated frog, which cannot of course feel, and cannot consciously perform, any movement. Yet if a drop of acid be placed on the lower surface of the thigh of a frog in this state, it will rub off the drop with the upper surface of the foot of the same leg. If this foot be cut off, it cannot thus act. "After some fruitless efforts, therefore, it gives up trying in that way, seems restless, as though, says Pflüger, it was seeking some other way, and at last it makes use of the foot of the other leg and succeeds in rubbing off the acid. Notably we have here not merely contractions of muscles, but combined and harmonized contractions in due sequence for a special purpose. These are actions that have all the appearance of being guided by intelligence and instigated by will in an animal, the recognized organ of whose intelligence and will has been removed." [10]

We see the difference between reflex and voluntary movements in very young children not being able to perform, as I am informed by Sir Henry Holland, certain acts somewhat analogous to those of sneezing and coughing, namely, in their not being able to blow their noses (i. e. to compress the nose and blow violently through the passage), and in their not being able to clear their throats of phlegm. They have to learn to perform these acts, yet they are performed by us, when a little older, almost as easily as reflex actions. Sneezing and coughing, however, can be controlled by the will only partially or not at all; whilst the clearing the throat

---

[10] Dr. Maudsley, 'Body and Mind,' 1870, p. 8.

and blowing the nose are completely under our command.

When we are conscious of the presence of an irritating particle in our nostrils or windpipe—that is, when the same sensory nerve-cells are excited, as in the case of sneezing and coughing—we can voluntarily expel the particle by forcibly driving air through these passages; but we cannot do this with nearly the same force, rapidity, and precision, as by a reflex action. In this latter case the sensory nerve-cells apparently excite the motor nerve-cells without any waste of power by first communicating with the cerebral hemispheres—the seat of our consciousness and volition. In all cases there seems to exist a profound antagonism between the same movements, as directed by the will and by a reflex stimulant, in the force with which they are performed and in the facility with which they are excited. As Claude Bernard asserts, "L'influence du cerveau tend donc à entraver les mouvements réflexes, à limiter leur force et leur étendue." [11]

The conscious wish to perform a reflex action sometimes stops or interrupts its performance, though the proper sensory nerves may be stimulated. For instance, many years ago I laid a small wager with a dozen young men that they would not sneeze if they took snuff, although they all declared that they invariably did so; accordingly they all took a pinch, but from wishing much to succeed, not one sneezed, though their eyes watered, and all, without exception, had to pay me the wager. Sir H. Holland remarks [12] that attention paid to the act of swallowing interferes with the proper movements; from which it probably follows,

---

[11] See the very interesting discussion on the whole subject by Claude Bernard, 'Tissus Vivants,' 1866, p. 353--356.

[12] 'Chapters on Mental Physiology,' 1858, p. 85.

at least in part, that some persons find it so difficult to swallow a pill.

Another familiar instance of a reflex action is the involuntary closing of the eyelids when the surface of the eye is touched. A similar winking movement is caused when a blow is directed towards the face; but this is an habitual and not a strictly reflex action, as the stimulus is conveyed through the mind and not by the excitement of a peripheral nerve. The whole body and head are generally at the same time drawn suddenly backwards. These latter movements, however, can be prevented, if the danger does not appear to the imagination imminent; but our reason telling us that there is no danger does not suffice. I may mention a trifling fact, illustrating this point, and which at the time amused me. I put my face close to the thick glass-plate in front of a puff-adder in the Zoological Gardens, with the firm determination of not starting back if the snake struck at me; but, as soon as the blow was struck, my resolution went for nothing, and I jumped a yard or two backwards with astonishing rapidity. My will and reason were powerless against the imagination of a danger which had never been experienced.

The violence of a start seems to depend partly on the vividness of the imagination, and partly on the condition, either habitual or temporary, of the nervous system. He who will attend to the starting of his horse, when tired and fresh, will perceive how perfect is the gradation from a mere glance at some unexpected object, with a momentary doubt whether it is dangerous, to a jump so rapid and violent, that the animal probably could not voluntarily whirl round in so rapid a manner. The nervous system of a fresh and highly-fed horse sends its order to the motory system so quickly, that no time is allowed for him to consider whether or not the danger is real. After one violent start, when he is excited and the blood flows freely through his brain, he is very apt to start again; and so it is, as I have noticed, with young infants.

A start from a sudden noise, when the stimulus is conveyed through the auditory nerves, is always accompanied in grown-up persons by the winking of the eyelids.[13] I observed, however, that though my infants started at sudden sounds, when under a fortnight old, they certainly did not always wink their eyes, and I believe never did so. The start of an older infant apparently represents a vague catching hold of something to prevent falling. I shook a pasteboard box close before the eyes of one of my infants, when 114 days old, and it did not in the least wink; but when I put a few comfits into the box, holding it in the same position as before, and rattled them, the child blinked its eyes violently every time, and started a little. It was obviously impossible that a carefully-guarded infant could have learnt by experience that a rattling sound near its eyes indicated danger to them. But such experience will have been slowly gained at a later age during a long series of generations; and from what we know of inheritance, there is nothing improbable in the transmission of a habit to the offspring at an earlier age than that at which it was first acquired by the parents.

From the foregoing remarks it seems probable that some actions, which were at first performed consciously, have become through habit and association converted into reflex actions, and are now so firmly fixed and inherited, that they are performed, even when not of the

[13] Müller remarks ('Elements of Physiology,' Eng. tr. vol. ii. p. 1311) on starting being always accompanied by the closure of the eyelids.

least use,[14] as often as the same causes arise, which originally excited them in us through the volition. In such cases the sensory nerve-cells excite the motor cells, without first communicating with those cells on which our consciousness and volition depend. It is probable that sneezing and coughing were originally acquired by the habit of expelling, as violently as possible, any irritating particle from the sensitive air-passages. As far as time is concerned, there has been more than enough for these habits to have become innate or converted into reflex actions; for they are common to most or all of the higher quadrupeds, and must therefore have been first acquired at a very remote period. Why the act of clearing the throat is not a reflex action, and has to be learnt by our children, I cannot pretend to say; but we can see why blowing the nose on a handkerchief has to be learnt.

It is scarcely credible that the movements of a headless frog, when it wipes off a drop of acid or other object from its thigh, and which movements are so well co-ordinated for a special purpose, were not at first performed voluntarily, being afterwards rendered easy through long-continued habit so as at last to be performed unconsciously, or independently of the cerebral hemispheres.

So again it appears probable that starting was originally acquired by the habit of jumping away as quickly as possible from danger, whenever any of our senses gave us warning. Starting, as we have seen, is accompanied by the blinking of the eyelids so as to protect the eyes, the most tender and sensitive organs

[14] Dr. Maudsley remarks ('Body and Mind,' p. 10) that "reflex movements which commonly effect a useful end may, under the changed circumstances of disease, do great mischief, becoming even the occasion of violent suffering and of a most painful death."

of the body; and it is, I believe, always accompanied by a sudden and forcible inspiration, which is the natural preparation for any violent effort. But when a man or horse starts, his heart beats wildly against his ribs, and here it may be truly said we have an organ which has never been under the control of the will, partaking in the general reflex movements of the body. To this point, however, I shall return in a future chapter.

The contraction of the iris, when the retina is stimulated by a bright light, is another instance of a movement, which it appears cannot possibly have been at first voluntarily performed and then fixed by habit; for the iris is not known to be under the conscious control of the will in any animal. In such cases some explanation, quite distinct from habit, will have to be discovered. The radiation of nerve-force from strongly-excited nerve-cells to other connected cells, as in the case of a bright light on the retina causing a sneeze, may perhaps aid us in understanding how some reflex actions originated. A radiation of nerve-force of this kind, if it caused a movement tending to lessen the primary irritation, as in the case of the contraction of the iris preventing too much light from falling on the retina, might afterwards have been taken advantage of and modified for this special purpose.

It further deserves notice that reflex actions are in all probability liable to slight variations, as are all corporeal structures and instincts; and any variations which were beneficial and of sufficient importance, would tend to be preserved and inherited. Thus reflex actions, when once gained for one purpose, might afterwards be modified independently of the will or habit, so as to serve for some distinct purpose. Such cases would be parallel with those which, as we have every reason to believe, have occurred with many instincts; for although some instincts have been developed simply through long-continued and inherited habit, other highly complex ones have been developed through the preservation of variations of pre-existing instincts—that is, through natural selection.

I have discussed at some little length, though as I am well aware, in a very imperfect manner, the acquirement of reflex actions, because they are often brought into play in connection with movements expressive of our emotions; and it was necessary to show that at least some of them might have been first acquired through the will in order to satisfy a desire, or to relieve a disagreeable sensation.

Associated habitual movements in the lower animals.—I have already given in the case of Man several instances of movements associated with various states of the mind or body, which are now purposeless, but which were originally of use, and are still of use under certain circumstances. As this subject is very important for us, I will here give a considerable number of analogous facts, with reference to animals; although many of them are of a very trifling nature. My object is to show that certain movements were originally performed for a definite end, and that, under nearly the same circumstances, they are still pertinaciously performed through habit when not of the least use. That the tendency in most of the following cases is inherited, we may infer from such actions being performed in the same manner by all the individuals, young and old, of the same species. We shall also see that they are excited by the most diversified, often circuitous, and sometimes mistaken associations.

Dogs, when they wish to go to sleep on a carpet or

other hard surface, generally turn round and round and scratch the ground with their fore-paws in a senseless manner, as if they intended to trample down the grass and scoop out a hollow, as no doubt their wild parents did, when they lived on open grassy plains or in the woods. Jackals, fennecs, and other allied animals in the Zoological Gardens, treat their straw in this manner; but it is a rather odd circumstance that the keepers, after observing for some months, have never seen the wolves thus behave. A semi-idiotic dog—and an animal in this condition would be particularly liable to follow a senseless habit—was observed by a friend to turn completely round on a carpet thirteen times before going to sleep.

Many carnivorous animals, as they crawl towards their prey and prepare to rush or spring on it, lower their heads and crouch, partly, as it would appear, to hide themselves, and partly to get ready for their rush; and this habit in an exaggerated form has become hereditary in our pointers and setters. Now I have noticed scores of times that when two strange dogs meet on an open road, the one which first sees the other, though at the distance of one or two hundred yards, after the first glance always lowers its head, generally crouches a little, or even lies down; that is, he takes the proper attitude for concealing himself and for making a rush or spring, although the road is quite open and the distance great. Again, dogs of all kinds when intently watching and slowly approaching their prey, frequently

FIG. 4.—Small dog watching a cat on a table. From a photograph taken by Mr. Rejlander.

keep one of their fore-legs doubled up for a long time, ready for the next cautious step; and this is eminently characteristic of the pointer. But from habit they behave in exactly the same manner whenever their attention is aroused (fig. 4). I have seen a dog at the foot of a high wall, listening attentively to a sound on the opposite side, with one leg doubled up; and in this case there could have been no intention of making a cautious approach.

Dogs after voiding their excrement often make with all four feet a few scratches backwards, even on a bare stone pavement, as if for the purpose of covering up their excrement with earth, in nearly the same manner as do cats. Wolves and jackals behave in the Zoological Gardens in exactly the same manner, yet, as I am assured by the keepers, neither wolves, jackals, nor foxes, when they have the means of doing so, ever cover up their excrement, any more than do dogs. All these animals, however, bury superfluous food. Hence, if we rightly understand the meaning of the above cat-like habit, of which there can be little doubt, we have a purposeless remnant of an habitual movement, which was originally followed by some remote progenitor of the dog-genus for a definite purpose, and which has been retained for a prodigious length of time.

Dogs and jackals[15] take much pleasure in rolling and rubbing their necks and backs on carrion. The odour seems delightful to them, though dogs at least do not eat carrion. Mr. Bartlett has observed wolves for me, and has given them carrion, but has never seen them roll on it. I have heard it remarked, and I believe it to be true, that the larger dogs, which are probably descended from wolves, do not so often roll in

---

[15] See Mr. F. H. Salvin's account of a tame jackal in 'Land and Water,' October, 1869.

carrion as do smaller dogs, which are probably descended from jackals. When a piece of brown biscuit is offered to a terrier of mine and she is not hungry (and I have heard of similar instances), she first tosses it about and worries it, as if it were a rat or other prey; she then repeatedly rolls on it precisely as if it were a piece of carrion, and at last eats it. It would appear that an imaginary relish has to be given to the distasteful morsel; and to effect this the dog acts in his habitual manner, as if the biscuit was a live animal or smelt like carrion, though he knows better than we do that this is not the case. I have seen this same terrier act in the same manner after killing a little bird or mouse.

Dogs scratch themselves by a rapid movement of one of their hind-feet; and when their backs are rubbed with a stick, so strong is the habit, that they cannot help rapidly scratching the air or the ground in a useless and ludicrous manner. The terrier just alluded to, when thus scratched with a stick, will sometimes show her delight by another habitual movement, namely, by licking the air as if it were my hand.

Horses scratch themselves by nibbling those parts of their bodies which they can reach with their teeth; but more commonly one horse shows another where he wants to be scratched, and they then nibble each other. A friend whose attention I had called to the subject, observed that when he rubbed his horse's neck, the animal protruded his head, uncovered his teeth, and moved his jaws, exactly as if nibbling another horse's neck, for he could never have nibbled his own neck. If a horse is much tickled, as when curry-combed, his wish to bite something becomes so intolerably strong, that he will clatter his teeth together, and though not vicious, bite his groom. At the same time from habit he closely depresses his ears, so as to protect them from being bitten, as if he were fighting with another horse.

A horse when eager to start on a journey makes the nearest approach which he can to the habitual movement of progression by pawing the ground. Now when horses in their stalls are about to be fed and are eager for their corn, they paw the pavement or the straw. Two of my horses thus behave when they see or hear the corn given to their neighbours. But here we have what may almost be called a true expression, as pawing the ground is universally recognized as a sign of eagerness.

Cats cover up their excrements of both kinds with earth; and my grandfather[17] saw a kitten scraping ashes over a spoonful of pure water spilt on the hearth; so that here an habitual or instinctive action was falsely excited, not by a previous act or by odour, but by eyesight. It is well known that cats dislike wetting their feet, owing, it is probable, to their having aboriginally inhabited the dry country of Egypt; and when they wet their feet they shake them violently. My daughter poured some water into a glass close to the head of a kitten; and it immediately shook its feet in the usual manner; so that here we have an habitual movement falsely excited by an associated sound instead of by the sense of touch.

Kittens, puppies, young pigs and probably many other young animals, alternately push with their fore-feet against the mammary glands of their mothers, to excite a freer secretion of milk, or to make it flow. Now it is very common with young cats, and not at all rare with old cats of the common and Persian breeds (be-

---

[16] Dr. Darwin, 'Zoonomia,' 1794, vol. i. p. 160. I find that the fact of cats protruding their feet when pleased is also noticed (p. 151) in this work.

lieved by some naturalists to be specifically extinct), when comfortably lying on a warm shawl or other soft substance, to pound it quietly and alternately with their fore-feet; their toes being spread out and claws slightly protruded, precisely as when sucking their mother. That it is the same movement is clearly shown by their often at the same time taking a bit of the shawl into their mouths and sucking it; generally closing their eyes and purring from delight. This curious movement is commonly excited only in association with the sensation of a warm soft surface; but I have seen an old cat, when pleased by having its back scratched, pounding the air with its feet in the same manner; so that this action has almost become the expression of a pleasurable sensation.

Having referred to the act of sucking, I may add that this complex movement, as well as the alternate protrusion of the fore-feet, are reflex actions; for they are performed if a finger moistened with milk is placed in the mouth of a puppy, the front part of whose brain has been removed.[17] It has recently been stated in France, that the action of sucking is excited solely through the sense of smell, so that if the olfactory nerves of a puppy are destroyed, it never sucks. In like manner the wonderful power which a chicken possesses only a few hours after being hatched, of picking up small particles of food, seems to be started into action through the sense of hearing; for with chickens hatched by artificial heat, a good observer found that "making a noise with the finger-nail against a board, in imitation of the hen-mother, first taught them to peck at their meat."[18]

[17] Carpenter, 'Principles of Comparative Physiology,' 1854, p. 690, and Müller's 'Elements of Physiology,' Eng. translat. vol. ii. p. 936.

[18] Mowbray on 'Poultry,' 6th edit. 1830, p. 54.

I will give only one other instance of an habitual and purposeless movement. The Sheldrake (*Tadorna*) feeds on the sands left uncovered by the tide, and when a worm-cast is discovered, "it begins patting the ground with its feet, dancing as it were, over the hole;" and this makes the worm come to the surface. Now Mr. St. John says, that when his tame Sheldrakes "came to ask for food, they patted the ground in an impatient and rapid manner."[19] This therefore may almost be considered as their expression of hunger. Mr. Bartlett informs me that the Flamingo and the Kagu (*Rhinochetus jubatus*) when anxious to be fed, beat the ground with their feet in the same odd manner. So again Kingfishers, when they catch a fish, always beat it until it is killed; and in the Zoological Gardens they always beat the raw meat, with which they are sometimes fed, before devouring it.

We have now, I think, sufficiently shown the truth of our first Principle, namely, that when any sensation, desire, dislike, &c., has led during a long series of generations to some voluntary movement, then a tendency to the performance of a similar movement will almost certainly be excited, whenever the same, or any analogous or associated sensation &c., although very weak, is experienced; notwithstanding that the movement in this case may not be of the least use. Such habitual movements are often, or generally inherited; and they then differ but little from reflex actions. When we treat of the special expressions of man, the latter part of our first Principle, as given at the commencement of this chapter, will be seen to hold good; namely, that when movements, associated through habit with certain states

[19] See the account given by this excellent observer in 'Wild Sports of the Highlands,' 1846, p. 142.

of the mind, are partially repressed by the will, the strictly involuntary muscles, as well as those which are least under the separate control of the will, are liable still to act; and their action is often highly expressive. Conversely, when the will is temporarily or permanently weakened, the voluntary muscles fail before the involuntary. It is a fact familiar to pathologists, as Sir C. Bell remarks,[20] "that when debility arises from affection of the brain, the influence is greatest on those muscles which are, in their natural condition, most under the command of the will." We shall, also, in our future chapters, consider another proposition included in our first Principle; namely, that the checking of one habitual movement sometimes requires other slight movements; these latter serving as a means of expression.

---

[20] 'Philosophical Translations,' 1823, p. 182.

CHAPTER II.

General Principles of Expression—*continued.*

The Principle of Antithesis—Instances in the dog and cat —Origin of the principle—Conventional signs—The principle of antithesis has not arisen from opposite actions being consciously performed under opposite impulses.

We will now consider our second Principle, that of Antithesis. Certain states of the mind lead, as we have seen in the last chapter, to certain habitual movements which were primarily, or may still be, of service; and we shall find that when a directly opposite state of mind is induced, there is a strong and involuntary tendency to the performance of movements of a directly opposite nature, though these have never been of any service. A few striking instances of antithesis will be given, when we treat of the special expressions of man; but as, in these cases, we are particularly liable to confound conventional or artificial gestures and expressions with those which are innate or universal, and which alone deserve to rank as true expressions, I will in the present chapter almost confine myself to the lower animals.

When a dog approaches a strange dog or man in a savage or hostile frame of mind he walks upright and very stiffly; his head is slightly raised, or not much lowered; the tail is held erect and quite rigid; the hairs bristle, especially along the neck and back; the pricked ears are directed forwards, and the eyes have a fixed stare: (see figs. 5 and 7). These actions, as will hereafter be explained, follow from the dog's intention to attack his enemy, and are thus to a large extent intelligible. As he prepares to spring with a savage growl on his enemy, the canine teeth are uncovered, and the ears are pressed close backwards on the head; but with these latter actions, we are not here concerned. Let us now suppose that the dog suddenly discovers that the man he is approaching, is not a stranger, but his master; and let it be observed how completely and instantaneously his whole bearing is reversed. Instead of walking upright, the body sinks downwards or even crouches, and is thrown into flexuous movements; his tail, instead of being held stiff and upright, is lowered and wagged from side to side; his hair instantly becomes smooth; his ears are depressed and drawn backwards, but not closely to the head; and his lips hang loosely. From the drawing back of the ears, the eyelids become elongated, and the eyes no longer appear round and staring. It should be added that the animal is at such times in an excited condition from joy; and nerve-force will be generated in excess, which naturally leads to action of some kind. Not one of the above movements, so clearly expressive of affection, are of the least direct service to the animal. They are explicable, as far as I can see, solely from being in complete opposition or antithesis to the attitude and movements which, from intelligible causes, are assumed when a dog intends to fight, and which consequently are expressive of anger. I request the reader to look at the four accompanying sketches, which have been given in order to recall vividly the appearance of a dog under these two states of mind. It is, however, not a little difficult to represent affection in a dog, whilst caressing his master and wagging his tail, as the essence of

FIG. 5.—Dog approaching another dog with hostile intentions. By Mr. Riviere.

FIG. 6.—The same in a humble and affectionate frame of mind. By Mr. Riviere.

FIG. 7.—Half-bred Shepherd Dog in the same state as in Fig. 5. By Mr. A. May.

FIG. 8.—The same caressing his master. By Mr. A. May.

the expression lies in the continuous flexuous movements.

We will now turn to the cat. When this animal is threatened by a dog, it arches its back in a surprising manner, erects its hair, opens its mouth and spits. But we are not here concerned with this well-known attitude, expressive of terror combined with anger; we are concerned only with that of rage or anger. This is not often seen, but may be observed when two cats are fighting together; and I have seen it well exhibited by a savage cat whilst plagued by a boy. The attitude is almost exactly the same as that of a tiger disturbed and growling over its food, which every one must have beheld in menageries. The animal assumes a crouching position, with the body extended; and the whole tail, or the tip alone, is lashed or curled from side to side. The hair is not in the least erect. Thus far, the attitude and movements are nearly the same as when the animal is prepared to spring on its prey, and when, no doubt, it feels savage. But when preparing to fight, there is this difference, that the ears are closely pressed backwards; the mouth is partially opened, showing the teeth; the fore feet are occasionally struck out with protruded claws; and the animal occasionally utters a fierce growl. (See figs. 9 and 10.) All, or almost all, these actions naturally follow (as hereafter to be explained), from the cat's manner and intention of attacking its enemy.

Let us now look at a cat in a directly opposite frame of mind, whilst feeling affectionate and caressing her master; and mark how opposite is her attitude in every respect. She now stands upright with her back slightly arched, which makes the hair appear rather rough, but it does not bristle; her tail, instead of being extended and lashed from side to side, is held quite stiff and perpendicularly upwards; her ears are erect and pointed; her mouth is closed; and she rubs against her master with a purr instead of a growl. Let it further be observed how widely different is the whole bearing of an affectionate cat from that of a dog, when with his body crouching and flexuous, his tail lowered and wagging, and ears depressed, he caresses his master. This contrast in the attitudes and movements of these two carnivorous animals, under the same pleased and affectionate frame of mind, can be explained, as it appears to me, solely by their movements standing in complete antithesis to those which are naturally assumed, when these animals feel savage and are prepared either to fight or to seize their prey.

In these cases of the dog and cat, there is every reason to believe that the gestures both of hostility and affection are innate or inherited; for they are almost identically the same in the different races of the species, and in all the individuals of the same race, both young and old.

I will here give one other instance of antithesis in expression. I formerly possessed a large dog, who, like every other dog, was much pleased to go out walking. He showed his pleasure by trotting gravely before me with high steps, head much raised, moderately erected ears, and tail carried aloft but not stiffly. Not far from my house a path branches off to the right, leading to the hot-house, which I used often to visit for a few moments, to look at my experimental plants. This was always a great disappointment to the dog, as he did not know whether I should continue my walk; and the instantaneous and complete change of expression which came over him as soon as my body swerved in the least towards the path (and I sometimes tried this as an experiment) was laughable. His look of dejection was

FIG. 9.—Cat, savage, and prepared to fight, drawn from life by Mr. Wood.

Fig. 10.—Cat in an affectionate frame of mind, by Mr. Wood.

known to every member of the family, and was called his *hot-house face.* This consisted in the head drooping much, the whole body sinking a little and remaining motionless; the ears and tail falling suddenly down, but the tail was by no means wagged. With the falling of the ears and of his great chaps, the eyes became much changed in appearance, and I fancied that they looked less bright. His aspect was that of piteous, hopeless dejection; and it was, as I have said, laughable, as the cause was so slight. Every detail in his attitude was in complete opposition to his former joyful yet dignified bearing; and can be explained, as it appears to me, in no other way, except through the principle of antithesis. Had not the change been so instantaneous, I should have attributed it to his lowered spirits affecting, as in the case of man, the nervous system and circulation, and consequently the tone of his whole muscular frame; and this may have been in part the cause.

We will now consider how the principle of antithesis in expression has arisen. With social animals, the power of intercommunication between the members of the same community,—and with other species, between the opposite sexes, as well as between the young and the old,—is of the highest importance to them. This is generally effected by means of the voice, but it is certain that gestures and expressions are to a certain extent mutually intelligible. Man not only uses inarticulate cries, gestures, and expressions, but has invented articulate language; if, indeed, the word *invented* can be applied to a process, completed by innumerable steps, half-consciously made. Any one who has watched monkeys will not doubt that they perfectly understand each other's gestures and expression, and to a large extent, as Rengger asserts,[1] those of man. An animal when going to attack another, or when afraid of another, often makes itself appear terrible, by erecting its hair, thus increasing the apparent bulk of its body, by showing its teeth, or brandishing its horns, or by uttering fierce sounds.

As the power of intercommunication is certainly of high service to many animals, there is no *à priori* improbability in the supposition, that gestures manifestly of an opposite nature to those by which certain feelings are already expressed, should at first have been voluntarily employed under the influence of an opposite state of feeling. The fact of the gestures being now innate, would be no valid objection to the belief that they were at first intentional; for if practised during many generations, they would probably at last be inherited. Nevertheless it is more than doubtful, as we shall immediately see, whether any of the cases which come under our present head of antithesis, have thus originated.

With conventional signs which are not innate, such as those used by the deaf and dumb and by savages, the principle of opposition or antithesis has been partially brought into play. The Cistercian monks thought it sinful to speak, and as they could not avoid holding some communication, they invented a gesture language, in which the principle of opposition seems to have been employed.[2] Dr. Scott, of the Exeter Deaf and Dumb Institution, writes to me that "opposites are greatly used in teaching the deaf and dumb, who have a lively sense of them." Nevertheless I have been surprised

---

[1] 'Naturgeschichte der Säugethiere von Paraguay,' 1830, s. 55.

[2] Mr. Tylor gives an account of the Cistercian gesture-language in his 'Early History of Mankind' (2nd edit. 1870, p. 40), and makes some remarks on the principle of opposition in gestures.

how few unequivocal instances can be adduced. This depends partly on all the signs having commonly had some natural origin; and partly on the practice of the deaf and dumb and of savages to contract their signs as much as possible for the sake of rapidity.[3] Hence their natural source or origin often becomes doubtful or is completely lost; as is likewise the case with articulate language.

Many signs, moreover, which plainly stand in opposition to each other, appear to have had on both sides a significant origin. This seems to hold good with the signs used by the deaf and dumb for light and darkness, for strength and weakness, &c. In a future chapter I shall endeavour to show that the opposite gestures of affirmation and negation, namely, vertically nodding and laterally shaking the head, have both probably had a natural beginning. The waving of the hand from right to left, which is used as a negative by some savages, may have been invented in imitation of shaking the head; but whether the opposite movement of waving the hand in a straight line from the face, which is used in affirmation, has arisen through antithesis or in some quite distinct manner, is doubtful.

If we now turn to the gestures which are innate or common to all the individuals of the same species, and which come under the present head of antithesis, it is extremely doubtful, whether any of them were at first deliberately invented and consciously performed. With mankind the best instance of a gesture standing in direct

[3] See on this subject Dr. W. R. Scott's interesting work, 'The Deaf and Dumb,' 2nd edit. 1870, p. 12. He says, "This contracting of natural gestures into much shorter gestures than the natural expression requires, is very common amongst the deaf and dumb. This contracted gesture is frequently so shortened as nearly to lose all semblance of the natural one, but to the deaf and dumb who use it, it still has the force of the original expression."

opposition to other movements, naturally assumed under an opposite frame of mind, is that of shrugging the shoulders. This expresses impotence or an apology,—something which cannot be done, or cannot be avoided. The gesture is sometimes used consciously and voluntarily, but it is extremely improbable that it was at first deliberately invented, and afterwards fixed by habit; for not only do young children sometimes shrug their shoulders under the above states of mind, but the movement is accompanied, as will be shown in a future chapter, by various subordinate movements, which not one man in a thousand is aware of, unless he has specially attended to the subject.

Dogs when approaching a strange dog, may find it useful to show by their movements that they are friendly, and do not wish to fight. When two young dogs in play are growling and biting each other's faces and legs, it is obvious that they mutually understand each other's gestures and manners. There seems, indeed, some degree of instinctive knowledge in puppies and kittens, that they must not use their sharp little teeth or claws too freely in their play, though this sometimes happens and a squeal is the result; otherwise they would often injure each other's eyes. When my terrier bites my hand in play, often snarling at the same time, if he bites too hard and I say *gently, gently*, he goes on biting, but answers me by a few wags of the tail, which seems to say "Never mind, it is all fun." Although dogs do thus express, and may wish to express, to other dogs and to man, that they are in a friendly state of mind, it is incredible that they could ever have deliberately thought of drawing back and depressing their ears, instead of holding them erect,—of lowering and wagging their tails, instead of keeping them stiff and upright, &c., because they knew that these movements stood in direct opposition to those assumed under an opposite and savage frame of mind.

Again, when a cat, or rather when some early progenitor of the species, from feeling affectionate first slightly arched its back, held its tail perpendicularly upwards and pricked its ears, can it be believed that the animal consciously wished thus to show that its frame of mind was directly the reverse of that, when from being ready to fight or to spring on its prey, it assumed a crouching attitude, curled its tail from side to side and depressed its ears? Even still less can I believe that my dog voluntarily put on his dejected attitude and "*hot-house face*," which formed so complete a contrast to his previous cheerful attitude and whole bearing. It cannot be supposed that he knew that I should understand his expression, and that he could thus soften my heart and make me give up visiting the hot-house.

Hence for the development of the movements which come under the present head, some other principle, distinct from the will and consciousness, must have intervened. This principle appears to be that every movement which we have voluntarily performed throughout our lives has required the action of certain muscles; and when we have performed a directly opposite movement, an opposite set of muscles has been habitually brought into play,—as in turning to the right or to the left, in pushing away or pulling an object towards us, and in lifting or lowering a weight. So strongly are our intentions and movements associated together, that if we eagerly wish an object to move in any direction, we can hardly avoid moving our bodies in the same direction, although we may be perfectly aware that this can have no influence. A good illustration of this fact has already been given in the Introduction, namely, in the grotesque movements of a young and eager billiard-player, whilst watching the course of his ball. A man or child in a passion, if he tells any one in a loud voice to begone, generally moves his arm as if to push him away, although the offender may not be standing near, and although there may be not the least need to explain by a gesture what is meant. On the other hand, if we eagerly desire some one to approach us closely, we act as if pulling him towards us; and so in innumerable other instances.

As the performance of ordinary movements of an opposite kind, under opposite impulses of the will, has become habitual in us and in the lower animals, so when actions of one kind have become firmly associated with any sensation or emotion, it appears natural that actions of a directly opposite kind, though of no use, should be unconsciously performed through habit and association, under the influence of a directly opposite sensation or emotion. On this principle alone can I understand how the gestures and expressions which come under the present head of antithesis have originated. If indeed they are serviceable to man or to any other animal, in aid of inarticulate cries or language, they will likewise be voluntarily employed, and the habit will thus be strengthened. But whether or not of service as a means of communication, the tendency to perform opposite movements under opposite sensations or emotions would, if we may judge by analogy, become hereditary through long practice; and there cannot be a doubt that several expressive movements due to the principle of antithesis are inherited.

CHAPTER III.

GENERAL PRINCIPLES OF EXPRESSION—*concluded.*

The principle of direct action of the excited nervous system on the body, independently of the will and in part of habit—Change of colour in the hair—Trembling of the muscles—Modified secretions—Perspiration—Expression of extreme pain—Of rage, great joy, and terror—Contrast between the emotions which cause and do not cause expressive movements—Exciting and depressing states of the mind—Summary.

WE now come to our third Principle, namely, that certain actions which we recognize as expressive of certain states of the mind, are the direct result of the constitution of the nervous system, and have been from the first independent of the will, and, to a large extent, of habit. When the sensorium is strongly excited nerve-force is generated in excess, and is transmitted in certain directions, dependent on the connection of the nerve-cells, and, as far as the muscular system is concerned, on the nature of the movements which have been habitually practised. Or the supply of nerve-force may, as it appears, be interrupted. Of course every movement which we make is determined by the constitution of the nervous system; but actions performed in obedience to the will, or through habit, or through the principle of antithesis, are here as far as possible excluded. Our present subject is very obscure, but, from its importance, must be discussed at some little length; and it is always advisable to perceive clearly our ignorance.

The most striking case, though a rare and abnormal one, which can be adduced of the direct influence of the nervous system, when strongly affected, on the body, is the loss of colour in the hair, which has occasionally been observed after extreme terror or grief. One authentic instance has been recorded, in the case of a man brought out for execution in India, in which the change of colour was so rapid that it was perceptible to the eye.[1]

Another good case is that of the trembling of the muscles, which is common to man and to many, or most, of the lower animals. Trembling is of no service, often of much disservice, and cannot have been at first acquired through the will, and then rendered habitual in association with any emotion. I am assured by an eminent authority that young children do not tremble, but go into convulsions under the circumstances which would induce excessive trembling in adults. Trembling is excited in different individuals in very different degrees, and by the most diversified causes,—by cold to the surface, before fever-fits, although the temperature of the body is then above the normal standard; in blood-poisoning, delirium tremens, and other diseases; by general failure of power in old age; by exhaustion after excessive fatigue; locally from severe injuries, such as burns; and, in an especial manner, by the passage of a catheter. Of all emotions, fear notoriously is the most apt to induce trembling; but so do occasionally great anger and joy. I remember once seeing a boy who had just shot his first snipe on the wing, and his hands

[1] See the interesting cases collected by M. G. Pouchet in the 'Revue des Deux Mondes,' January 1, 1872, p. 79. An instance was also brought some years ago before the British Association at Belfast.

trembled to such a degree from delight, that he could not for some time reload his gun; and I have heard of an exactly similar case with an Australian savage, to whom a gun had been lent. Fine music, from the vague emotions thus excited, causes a shiver to run down the backs of some persons. There seems to be very little in common in the above several physical causes and emotions to account for trembling; and Sir J. Paget, to whom I am indebted for several of the above statements, informs me that the subject is a very obscure one. As trembling is sometimes caused by rage, long before exhaustion can have set in, and as it sometimes accompanies great joy, it would appear that any strong excitement of the nervous system interrupts the steady flow of nerve-force to the muscles.[2]

The manner in which the secretions of the alimentary canal and of certain glands—as the liver, kidneys, or mammæ—are affected by strong emotions, is another excellent instance of the direct action of the sensorium on these organs, independently of the will or of any serviceable associated habit. There is the greatest difference in different persons in the parts which are thus affected, and in the degree of their affection.

The heart, which goes on uninterruptedly beating night and day in so wonderful a manner, is extremely sensitive to external stimulants. The great physiologist, Claude Bernard,[3] has shown how the least excitement of a sensitive nerve reacts on the heart; even when a nerve is touched so slightly that no pain can possibly

---

[2] Müller remarks ('Elements of Physiology,' Eng. translat. vol. ii. p. 934) that when the feelings are very intense, "all the spinal nerves become affected to the extent of imperfect paralysis, or the excitement of trembling of the whole body."

[3] 'Leçons sur les Prop. des Tissus Vivants,' 1866, pp. 457--466.

be felt by the animal under experiment. Hence when the mind is strongly excited, we might expect that it would instantly affect in a direct manner the heart; and this is universally acknowledged and felt to be the case. Claude Bernard also repeatedly insists, and this deserves especial notice, that when the heart is affected it reacts on the brain; and the state of the brain again reacts through the pneumo-gastric nerve on the heart; so that under any excitement there will be much mutual action and reaction between these, the two most important organs of the body.

The vaso-motor system, which regulates the diameter of the small arteries, is directly acted on by the sensorium, as we see when a man blushes from shame; but in this latter case the checked transmission of nerve-force to the vessels of the face can, I think, be partly explained in a curious manner through habit. We shall also be able to throw some light, though very little, on the involuntary erection of the hair under the emotions of terror and rage. The secretion of tears depends, no doubt, on the connection of certain nerve-cells; but here again we can trace some few of the steps by which the flow of nerve-force through the requisite channels has become habitual under certain emotions.

A brief consideration of the outward signs of some of the stronger sensations and emotions will best serve to show us, although vaguely, in how complex a manner the principle under consideration of the direct action of the excited nervous system of the body, is combined with the principle of habitually associated, serviceable movements.

When animals suffer from an agony of pain, they generally writhe about with frightful contortions; and those which habitually use their voices utter piercing cries or groans. Almost every muscle of the body is brought into strong action. With man the mouth may be closely compressed, or more commonly the lips are retracted, with the teeth clenched or ground together. There is said to be "gnashing of teeth" in hell; and I have plainly heard the grinding of the molar teeth of a cow which was suffering acutely from inflammation of the bowels. The female hippopotamus in the Zoological Gardens, when she produced her young, suffered greatly; she incessantly walked about, or rolled on her sides, opening and closing her jaws, and clattering her teeth together.[4] With man the eyes stare wildly as in horrified astonishment, or the brows are heavily contracted. Perspiration bathes the body, and drops trickle down the face. The circulation and respiration are much affected. Hence the nostrils are generally dilated and often quiver; or the breath may be held until the blood stagnates in the purple face. If the agony be severe and prolonged, these signs all change; utter prostration follows, with fainting or convulsions.

A sensitive nerve when irritated transmits some influence to the nerve-cell, whence it proceeds; and this transmits its influence, first to the corresponding nerve-cell on the opposite side of the body, and then upwards and downwards along the cerebro-spinal column to other nerve-cells, to a greater or less extent, according to the strength of the excitement; so that, ultimately, the whole nervous system may be affected.[5] This involuntary transmission of nerve-force may or may not be accompa-

---

[4] Mr. Bartlett, "Notes on the Birth of a Hippopotamus," Proc. Zoolog. Soc. 1871, p. 255.

[5] See, on this subject, Claude Bernard, 'Tissus Vivants,' 1866, pp. 316, 337, 358. Virchow expresses himself to almost exactly the same effect in his essay "Ueber das Rückenmark" (Sammlung wissenschaft. Vorträge, 1871, s. 28).

nied by consciousness. Why the irritation of a nerve-cell should generate or liberate nerve-force is not known; but that this is the case seems to be the conclusion arrived at by all the greatest physiologists, such as Müller, Virchow, Bernard, &c.[6] As Mr. Herbert Spencer remarks, it may be received as an "unquestionable truth that, at any moment, the existing quantity of liberated nerve-force, which in an inscrutable way produces in us the state we call feeling, *must* expend itself in some direction—*must* generate an equivalent manifestation of force somewhere;" so that, when the cerebro-spinal system is highly excited and nerve-force is liberated in excess, it may be expended in intense sensations, active thought, violent movements, or increased activity of the glands.[7] Mr. Spencer further maintains that an "overflow of nerve-force, undirected by any motive, will manifestly take the most habitual routes; and, if these do not suffice, will next overflow into the less habitual ones." Consequently the facial and respiratory muscles, which are the most used, will be apt to be first brought into action; then those of the upper extremities, next those of the lower, and finally those of the whole body.[8]

An emotion may be very strong, but it will have little tendency to induce movements of any kind, if it

---

[6] Müller ('Elements of Physiology,' Eng. translat. vol. ii. p. 932) in speaking of the nerves, says, "any sudden change of condition of whatever kind sets the nervous principle into action." See Virchow and Bernard on the same subject in passages in the two works referred to in my last foot-note.

[7] H. Spencer, 'Essays, Scientific, Political,' &c., Second Series, 1863, pp. 109, 111.

[8] Sir H. Holland, in speaking ('Medical Notes and Reflexions,' 1839, p. 328) of that curious state of body called the *fidgets*, remarks that it seems due to "an accumulation of some cause of irritation which requires muscular action for its relief."

has not commonly led to voluntary action for its relief or gratification; and when movements are excited, their nature is, to a large extent, determined by those which have often and voluntarily been performed for some definite end under the same emotion. Great pain urges all animals, and has urged them during endless generations, to make the most violent and diversified efforts to escape from the cause of suffering. Even when a limb or other separate part of the body is hurt, we often see a tendency to shake it, as if to shake off the cause, though this may obviously be impossible. Thus a habit of exerting with the utmost force all the muscles will have been established, whenever great suffering is experienced. As the muscles of the chest and vocal organs are habitually used, these will be particularly liable to be acted on, and loud, harsh screams or cries will be uttered. But the advantage derived from outcries has here probably come into play in an important manner; for the young of most animals, when in distress or danger, call loudly to their parents for aid, as do the members of the same community for mutual aid.

Another principle, namely, the internal consciousness that the power or capacity of the nervous system is limited, will have strengthened, though in a subordinate degree, the tendency to violent action under extreme suffering. A man cannot think deeply and exert his utmost muscular force. As Hippocrates long ago observed, if two pains are felt at the same time, the severer one dulls the other. Martyrs, in the ecstasy of their religious fervour have often, as it would appear, been insensible to the most horrid tortures. Sailors who are going to be flogged sometimes take a piece of lead into their mouths, in order to bite it with their utmost force, and thus to bear the pain. Parturient women prepare to exert their muscles to the utmost in order to relieve their sufferings.

We thus see that the undirected radiation of nerve-force from the nerve-cells which are first affected—the long-continued habit of attempting by struggling to escape from the cause of suffering—and the consciousness that voluntary muscular exertion relieves pain, have all probably concurred in giving a tendency to the most violent, almost convulsive, movements under extreme suffering; and such movements, including those of the vocal organs, are universally recognized as highly expressive of this condition.

As the mere touching of a sensitive nerve reacts in a direct manner on the heart, severe pain will obviously react on it in like manner, but far more energetically. Nevertheless, even in this case, we must not overlook the indirect effects of habit on the heart, as we shall see when we consider the signs of rage.

When a man suffers from an agony of pain, the perspiration often trickles down his face; and I have been assured by a veterinary surgeon that he has frequently seen drops falling from the belly and running down the inside of the thighs of horses, and from the bodies of cattle, when thus suffering. He has observed this, when there has been no struggling which would account for the perspiration. The whole body of the female hippopotamus, before alluded to, was covered with red-coloured perspiration whilst giving birth to her young. So it is with extreme fear; the same veterinary has often seen horses sweating from this cause; as has Mr. Bartlett with the rhinoceros; and with man it is a well-known symptom. The cause of perspiration bursting forth in these cases is quite obscure; but it is thought by some physiologists to be connected with the failing power of the capillary circulation; and we know that the vaso-motor system, which regulates the capillary circulation, is much influenced by the mind. With respect to the movements of certain muscles of the face under great suffering, as well as from other emotions, these will be best considered when we treat of the special expressions of man and of the lower animals.

We will now turn to the characteristic symptoms of Rage. Under this powerful emotion the action of the heart is much accelerated,[9] or it may be much disturbed. The face reddens, or it becomes purple from the impeded return of the blood, or may turn deadly pale. The respiration is laboured, the chest heaves, and the dilated nostrils quiver. The whole body often trembles. The voice is affected. The teeth are clenched or ground together, and the muscular system is commonly stimulated to violent, almost frantic action. But the gestures of a man in this state usually differ from the purposeless writhings and struggles of one suffering from an agony of pain; for they represent more or less plainly the act of striking or fighting with an enemy.

All these signs of rage are probably in large part, and some of them appear to be wholly, due to the direct action of the excited sensorium. But animals of all kinds, and their progenitors before them, when attacked or threatened by an enemy, have exerted their utmost powers in fighting and in defending themselves. Unless an animal does thus act, or has the intention, or at least the desire, to attack its enemy, it cannot properly be said to be enraged. An inherited habit of muscular exertion will thus have been gained in association with rage; and this will directly or indirectly affect vari-

---

[9] I am much indebted to Mr. A. H. Garrod for having informed me of M. Lorain's work on the pulse, in which a sphygmogram of a woman in a rage is given; and this shows much difference in the rate and other characters from that of the same woman in her ordinary state.

ous organs, in nearly the same manner as does great bodily suffering.

The heart no doubt will likewise be affected in a direct manner; but it will also in all probability be affected through habit; and all the more so from not being under the control of the will. We know that any great exertion which we voluntarily make, affects the heart, through mechanical and other principles which need not here be considered; and it was shown in the first chapter that nerve-force flows readily through habitually used channels,—through the nerves of voluntary or involuntary movement, and through those of sensation. Thus even a moderate amount of exertion will tend to act on the heart; and on the principle of association, of which so many instances have been given, we may feel nearly sure that any sensation or emotion, as great pain or rage, which has habitually led to much muscular action, will immediately influence the flow of nerve-force to the heart, although there may not be at the time any muscular exertion.

The heart, as I have said, will be all the more readily affected through habitual associations, as it is not under the control of the will. A man when moderately angry, or even when enraged, may command the movements of his body, but he cannot prevent his heart from beating rapidly. His chest will perhaps give a few heaves, and his nostrils just quiver, for the movements of respiration are only in part voluntary. In like manner those muscles of the face which are least obedient to the will, will sometimes alone betray a slight and passing emotion. The glands again are wholly independent of the will, and a man suffering from grief may command his features, but cannot always prevent the tears from coming into his eyes. A hungry man, if tempting food is placed before him, may not show his hunger by any outward gesture, but he cannot check the secretion of saliva.

Under a transport of Joy or of vivid Pleasure, there is a strong tendency to various purposeless movements, and to the utterance of various sounds. We see this in our young children, in their loud laughter, clapping of hands, and jumping for joy; in the bounding and barking of a dog when going out to walk with his master; and in the frisking of a horse when turned out into an open field. Joy quickens the circulation, and this stimulates the brain, which again reacts on the whole body. The above purposeless movements and increased heart-action may be attributed in chief part to the excited state of the sensorium,[10] and to the consequent undirected overflow, as Mr. Herbert Spencer insists, of nerve-force. It deserves notice, that it is chiefly the anticipation of a pleasure, and not its actual enjoyment, which leads to purposeless and extravagant movements of the body, and to the utterance of various sounds. We see this in our children when they expect any great pleasure or treat; and dogs, which have been bounding about at

---

[10] How powerfully intense joy excites the brain, and how the brain reacts on the body, is well shown in the rare cases of Psychical Intoxication. Dr. J. Crichton Browne ('Medical Mirror,' 1865) records the case of a young man of strongly nervous temperament, who, on hearing by a telegram that a fortune had been bequeathed him, first became pale, then exhilarated, and soon in the highest spirits, but flushed and very restless. He then took a walk with a friend for the sake of tranquillising himself, but returned staggering in his gait, uproariously laughing, yet irritable in temper, incessantly talking, and singing loudly in the public streets. It was positively ascertained that he had not touched any spirituous liquor, though every one thought that he was intoxicated. Vomiting after a time came on, and the half-digested contents of his stomach were examined, but no odour of alcohol could be detected. He then slept heavily, and on awaking was well, except that he suffered from headache, nausea, and prostration of strength.

the sight of a plate of food, when they get it do not show their delight by any outward sign, not even by wagging their tails. Now with animals of all kinds, the acquirement of almost all their pleasures, with the exception of those of warmth and rest, are associated, and have long been associated with active movements, as in the hunting or search for food, and in their courtship. Moreover, the mere exertion of the muscles after long rest or confinement is in itself a pleasure, as we ourselves feel, and as we see in the play of young animals. Therefore on this latter principle alone we might perhaps expect, that vivid pleasure would be apt to show itself conversely in muscular movements.

With all or almost all animals, even with birds, Terror causes the body to tremble. The skin becomes pale, sweat breaks out, and the hair bristles. The secretions of the alimentary canal and of the kidneys are increased, and they are involuntarily voided, owing to the relaxation of the sphincter muscles, as is known to be the case with man, and as I have seen with cattle, dogs, cats, and monkeys. The breathing is hurried. The heart beats quickly, wildly, and violently; but whether it pumps the blood more efficiently through the body may be doubted, for the surface seems bloodless and the strength of the muscles soon fails. In a frightened horse I have felt through the saddle the beating of the heart so plainly that I could have counted the beats. The mental faculties are much disturbed. Utter prostration soon follows, and even fainting. A terrified canary-bird has been seen not only to tremble and to turn white about the base of the bill, but to faint;[11] and I once caught a robin in a room, which fainted so completely, that for a time I thought it dead.

---

[11] Dr. Darwin, 'Zoonomia,' 1794, vol. i. p. 148.

Most of these symptoms are probably the direct result, independently of habit, of the disturbed state of the sensorium; but it is doubtful whether they ought to be wholly thus accounted for. When an animal is alarmed it almost always stands motionless for a moment, in order to collect its senses and to ascertain the source of danger, and sometimes for the sake of escaping detection. But headlong flight soon follows, with no husbanding of the strength as in fighting, and the animal continues to fly as long as the danger lasts, until utter prostration, with failing respiration and circulation, with all the muscles quivering and profuse sweating, renders further flight impossible. Hence it does not seem improbable that the principle of associated habit may in part account for, or at least augment, some of the abovenamed characteristic symptoms of extreme terror.

That the principle of associated habit has played an important part in causing the movements expressive of the foregoing several strong emotions and sensations, we may, I think, conclude from considering firstly, some other strong emotions which do not ordinarily require for their relief or gratification any voluntary movement; and secondly the contrast in nature between the so-called exciting and depressing states of the mind. No emotion is stronger than maternal love; but a mother may feel the deepest love for her helpless infant, and yet not show it by any outward sign; or only by slight caressing movements, with a gentle smile and tender eyes. But let any one intentionally injure her infant, and see what a change! how she starts up with threatening aspect, how her eyes sparkle and her face reddens, how her bosom heaves, nostrils dilate, and heart beats; for anger, and not maternal love, has habitually led to action. The love between the opposite sexes is widely different from maternal love; and when lovers meet, we know that their hearts beat quickly, their breathing is hurried, and their faces flush; for this love is not inactive like that of a mother for her infant.

A man may have his mind filled with the blackest hatred or suspicion, or be corroded with envy or jealousy, but as these feelings do not at once lead to action, and as they commonly last for some time, they are not shown by any outward sign, excepting that a man in this state assuredly does not appear cheerful or good-tempered. If indeed these feelings break out into overt acts, rage takes their place, and will be plainly exhibited. Painters can hardly portray suspicion, jealousy, envy, &c., except by the aid of accessories which tell the tale; and poets use such vague and fanciful expressions as "green-eyed jealousy." Spenser describes suspicion as "Foul, ill-favoured, and grim, under his eyebrows looking still askance," &c.; Shakespeare speaks of envy "as lean-faced in her loathsome case;" and in another place he says, "no black envy shall make my grave;" and again as "above pale envy's threatening reach."

Emotions and sensations have often been classed as exciting or depressing. When all the organs of the body and mind,—those of voluntary and involuntary movement, of perception, sensation, thought, &c.,—perform their functions more energetically and rapidly than usual, a man or animal may be said to be excited, and, under an opposite state, to be depressed. Anger and joy are from the first exciting emotions, and they naturally lead, more especially the former, to energetic movements, which react on the heart and this again on the brain. A physician once remarked to me as a proof of the exciting nature of anger, that a man when excessively jaded will sometimes invent imaginary offences and put himself into a passion, unconsciously for the sake of reinvigorating himself; and since hearing this remark, I have occasionally recognized its full truth.

Several other states of mind appear to be at first exciting, but soon become depressing to an extreme degree. When a mother suddenly loses her child, sometimes she is frantic with grief, and must be considered to be in an excited state; she walks wildly about, tears her hair or clothes, and wrings her hands. This latter action is perhaps due to the principle of antithesis, betraying an inward sense of helplessness and that nothing can be done. The other wild and violent movements may be in part explained by the relief experienced through muscular exertion, and in part by the undirected overflow of nerve-force from the excited sensorium. But under the sudden loss of a beloved person, one of the first and commonest thoughts which occurs, is that something more might have been done to save the lost one. An excellent observer,[12] in describing the behaviour of a girl at the sudden death of her father, says she "went about the house wringing her hands like a creature demented, saying 'It was her fault;' 'I should never have left him;' 'If I had only sat up with him,'" &c. With such ideas vividly present before the mind, there would arise, through the principle of associated habit, the strongest tendency to energetic action of some kind.

As soon as the sufferer is fully conscious that nothing can be done, despair or deep sorrow takes the place of frantic grief. The sufferer sits motionless, or gently rocks to and fro; the circulation becomes languid; respiration is almost forgotten, and deep sighs are drawn.

---

[12] Mrs. Oliphant, in her novel of 'Miss Majoribanks,' p. 362.

All this reacts on the brain, and prostration soon follows with collapsed muscles and dulled eyes. As associated habit no longer prompts the sufferer to action, he is urged by his friends to voluntary exertion, and not to give way to silent, motionless grief. Exertion stimulates the heart, and this reacts on the brain, and aids the mind to bear its heavy load.

Pain, if severe, soon induces extreme depression or prostration; but it is at first a stimulant and excites to action, as we see when we whip a horse, and as is shown by the horrid tortures inflicted in foreign lands on exhausted dray-bullocks, to rouse them to renewed exertion. Fear again is the most depressing of all the emotions; and it soon induces utter, helpless prostration, as if in consequence of, or in association with, the most violent and prolonged attempts to escape from the danger, though no such attempts have actually been made. Nevertheless, even extreme fear often acts at first as a powerful stimulant. A man or animal driven through terror to desperation, is endowed with wonderful strength, and is notoriously dangerous in the highest degree.

On the whole we may conclude that the principle of the direct action of the sensorium on the body, due to the constitution of the nervous system, and from the first independent of the will, has been highly influential in determining many expressions. Good instances are afforded by the trembling of the muscles, the sweating of the skin, the modified secretions of the alimentary canal and glands, under various emotions and sensations. But actions of this kind are often combined with others, which follow from our first principle, namely, that actions which have often been of direct or indirect service, under certain states of the mind, in order to gratify or relieve certain sensations, desires, &c., are still performed under analogous circumstances through mere habit although of no service. We have combinations of this kind, at least in part, in the frantic gestures of rage and in the writhings of extreme pain; and, perhaps, in the increased action of the heart and of the respiratory organs. Even when these and other emotions or sensations are aroused in a very feeble manner, there will still be a tendency to similar actions, owing to the force of long-associated habit; and those actions which are least under voluntary control will generally be longest retained. Our second principle of antithesis has likewise occasionally come into play.

Finally, so many expressive movements can be explained, as I trust will be seen in the course of this volume, through the three principles which have now been discussed, that we may hope hereafter to see all thus explained, or by closely analogous principles. It is, however, often impossible to decide how much weight ought to be attributed, in each particular case, to one of our principles, and how much to another; and very many points in the theory of Expression remain inexplicable.

CHAPTER IV.

MEANS OF EXPRESSION IN ANIMALS.

The emission of sounds—Vocal sounds—Sounds otherwise produced—Erection of the dermal appendages, hairs, feathers, &c., under the emotions of anger and terror—The drawing back of the ears as a preparation for fighting, and as an expression of anger—Erection of the ears and raising the head, a sign of attention.

IN this and the following chapter I will describe, but only in sufficient detail to illustrate my subject, the expressive movements, under different states of the mind, of some few well-known animals. But before considering them in due succession, it will save much useless repetition to discuss certain means of expression common to most of them.

The emission of Sounds.—With many kinds of animals, man included, the vocal organs are efficient in the highest degree as a means of expression. We have seen, in the last chapter, that when the sensorium is strongly excited, the muscles of the body are generally thrown into violent action; and as a consequence, loud sounds are uttered, however silent the animal may generally be, and although the sounds may be of no use. Hares and rabbits for instance, never, I believe, use their vocal organs except in the extremity of suffering; as, when a wounded hare is killed by the sportsman, or when a young rabbit is caught by a stoat. Cattle and horses suffer great pain in silence; but when this is excessive, and especially when associated with terror, they utter fearful sounds. I have often recognized, from a distance on the Pampas, the agonized death-bellow of the cattle, when caught by the lasso and hamstrung. It is said that horses, when attacked by wolves, utter loud and peculiar screams of distress.

Involuntary and purposeless contractions of the muscles of the chest and glottis, excited in the above manner, may have first given rise to the emission of vocal sounds. But the voice is now largely used by many animals for various purposes; and habit seems to have played an important part in its employment under other circumstances. Naturalists have remarked, I believe with truth, that social animals, from habitually using their vocal organs as a means of intercommunication, use them on other occasions much more freely than other animals. But there are marked exceptions to this rule, for instance, with the rabbit. The principle, also, of association, which is so widely extended in its power, has likewise played its part. Hence it follows that the voice, from having been habitually employed as a serviceable aid under certain conditions, inducing pleasure, pain, rage, &c., is commonly used whenever the same sensations or emotions are excited, under quite different conditions, or in a lesser degree.

The sexes of many animals incessantly call for each other during the breeding-season; and in not a few cases, the male endeavours thus to charm or excite the female. This, indeed, seems to have been the primeval use and means of development of the voice, as I have attempted to show in my 'Descent of Man.' Thus the use of the vocal organs will have become associated with the anticipation of the strongest pleasure which animals are capable of feeling. Animals which live in society often call to each other when separated, and evidently feel much joy at meeting; as we see with a horse, on the return of his companion, for whom he has been neighing. The mother calls incessantly for her lost young ones; for instance, a cow for her calf; and the young of many animals call for their mothers. When a flock of sheep is scattered, the ewes bleat incessantly for their lambs, and their mutual pleasure at coming together is manifest. Woe betide the man who meddles with the young of the larger and fiercer quadrupeds, if they hear the cry of distress from their young. Rage leads to the violent exertion of all the muscles, including those of the voice; and some animals, when enraged, endeavour to strike terror into their enemies by its power and harshness, as the lion does by roaring, and the dog by growling. I infer that their object is to strike terror, because the lion at the same time erects the hair of its mane, and the dog the hair along its back, and thus they make themselves appear as large and terrible as possible. Rival males try to excel and challenge each other by their voices, and this leads to deadly contests. Thus the use of the voice will have become associated with the emotion of anger, however it may be aroused. We have also seen that intense pain, like rage, leads to violent outcries, and the exertion of screaming by itself gives some relief; and thus the use of the voice will have become associated with suffering of any kind.

The cause of widely different sounds being uttered under different emotions and sensations is a very obscure subject. Nor does the rule always hold good that there is any marked difference. For instance with the dog, the bark of anger and that of joy do not differ much, though they can be distinguished. It is not probable that any precise explanation of the cause or source of each particular sound, under different states of the mind, will ever be given. We know that some animals, after being domesticated, have acquired the habit of uttering sounds which were not natural to them.[1] Thus domestic dogs, and even tamed jackals, have learnt to bark, which is a noise not proper to any species of the genus, with the exception of the *Canis latrans* of North America, which is said to bark. Some breeds, also, of the domestic pigeon have learnt to coo in a new and quite peculiar manner.

The character of the human voice, under the influence of various emotions, has been discussed by Mr. Herbert Spencer[2] in his interesting essay on Music. He clearly shows that the voice alters much under different conditions, in loudness and in quality, that is, in resonance and *timbre,* in pitch and intervals. No one can listen to an eloquent orator or preacher, or to a man calling angrily to another, or to one expressing astonishment, without being struck with the truth of Mr. Spencer's remarks. It is curious how early in life the modulation of the voice becomes expressive. With one of my children, under the age of two years, I clearly perceived that his humph of assent was rendered by a slight modulation strongly emphatic; and that by a peculiar whine his negative expressed obstinate determination. Mr. Spencer further shows that emotional speech, in all the above respects is intimately related to vocal music, and consequently to instrumental music; and he attempts to explain the characteristic qualities of both on physiological grounds—namely, on "the general law that a feeling is a stimulus to muscular action." It may be

---

[1] See the evidence on this head in my 'Variation of Animals and Plants under Domestication,' vol. i. p. 27. On the cooing of pigeons, vol. i. pp. 154, 155.

[2] 'Essays, Scientific, Political, and Speculative,' 1858. 'The Origin and Function of Music,' p. 359.

admitted that the voice is affected through this law; but the explanation appears to me too general and vague to throw much light on the various differences, with the exception of that of loudness, between ordinary speech and emotional speech, or singing.

This remark holds good, whether we believe that the various qualities of the voice originated in speaking under the excitement of strong feelings, and that these qualities have subsequently been transferred to vocal music; or whether we believe, as I maintain, that the habit of uttering musical sounds was first developed, as a means of courtship, in the early progenitors of man, and thus became associated with the strongest emotions of which they were capable,—namely, ardent love, rivalry and triumph. That animals utter musical notes is familiar to every one, as we may daily hear in the singing of birds. It is a more remarkable fact that an ape, one of the Gibbons, produces an exact octave of musical sounds, ascending and descending the scale by half-tones; so that this monkey "alone of brute mammals may be said to sing."[3] From this fact, and from the analogy of other animals, I have been led to infer that the progenitors of man probably uttered musical tones, before they had acquired the power of articulate speech; and that consequently, when the voice is used under any strong emotion, it tends to assume, through the principle of association, a musical character. We can plainly perceive, with some of the lower animals, that the males employ their voices to please the females, and that they

[3] 'The Descent of Man,' 1870, vol. ii. p. 332. The words quoted are from Professor Owen. It has lately been shown that some quadrupeds much lower in the scale than monkeys, namely Rodents, are able to produce correct musical tones: see the account of a singing Hesperomys, by the Rev. S. Lockwood, in the 'American Naturalist,' vol. v. December, 1871, p. 761.

themselves take pleasure in their own vocal utterances; but why particular sounds are uttered, and why these give pleasure cannot at present be explained.

That the pitch of the voice bears some relation to certain states of feeling is tolerably clear. A person gently complaining of ill-treatment, or slightly suffering, almost always speaks in a high-pitched voice. Dogs, when a little impatient, often make a high piping note through their noses, which at once strikes us as plaintive;[4] but how difficult it is to know whether the sound is essentially plaintive, or only appears so in this particular case, from our having learnt by experience what it means! Rengger, states[5] that the monkeys (*Cebus azaræ*), which he kept in Paraguay, expressed astonishment by a half-piping, half-snarling noise; anger or impatience, by repeating the sound *hu hu* in a deeper, grunting voice; and fright or pain, by shrill screams. On the other hand, with mankind, deep groans and high piercing screams equally express an agony of pain. Laughter may be either high or low; so that, with adult men, as Haller long ago remarked,[6] the sound partakes of the character of the vowels (as pronounced in German) *O* and *A*; whilst with children and women, it has more of the character of *E* and *I*; and these latter vowel-sounds naturally have, as Helmholtz has shown, a higher pitch than the former; yet both tones of laughter equally express enjoyment or amusement.

In considering the mode in which vocal utterances express emotion, we are naturally led to inquire into

---

[4] Mr. Tylor ('Primitive Culture,' 1871, vol. i. p. 166), in his discussion on this subject, alludes to the whining of the dog.

[5] 'Naturgeschichte der Säugethiere von Paraguay,' 1830, s. 46.

[6] Quoted by Gratiolet, 'De la Physionomie,' 1865, p. 115.

the cause of what is called "expression" in music. Upon this point Mr. Litchfield, who has long attended to the subject of music, has been so kind as to give me the following remarks:—"The question, what is the essence of musical 'expression' involves a number of obscure points, which, so far as I am aware, are as yet unsolved enigmas. Up to a certain point, however, any law which is found to hold as to the expression of the emotions by simple sounds must apply to the more developed mode of expression in song, which may be taken as the primary type of all music. A great part of the emotional effect of a song depends on the character of the action by which the sounds are produced. In songs, for instance, which express great vehemence of passion, the effect often chiefly depends on the forcible utterance of some one or two characteristic passages which demand great exertion of vocal force; and it will be frequently noticed that a song of this character fails of its proper effect when sung by a voice of sufficient power and range to give the characteristic passages without much exertion. This is, no doubt, the secret of the loss of effect so often produced by the transposition of a song from one key to another. The effect is thus seen to depend not merely on the actual sounds, but also in part on the nature of the action which produces the sounds. Indeed it is obvious that whenever we feel the 'expression' of a song to be due to its quickness or slowness of movement—to smoothness of flow, loudness of utterance, and so on—we are, in fact, interpreting the muscular actions which produce sound, in the same way in which we interpret muscular action generally. But this leaves unexplained the more subtle and more specific effect which we call the *musical* expression of the song—the delight given by its melody, or even by the separate sounds which make up the melody. This is an effect indefinable in language—one which, so far as I am aware, no one has been able to analyse, and which the ingenious speculation of Mr. Herbert Spencer as to the origin of music leaves quite unexplained. For it is certain that the *melodic* effect of a series of sounds does not depend in the least on their loudness or softness, or on their *absolute* pitch. A tune is always the same tune, whether it is sung loudly or softly, by a child or a man; whether it is played on a flute or on a trombone. The purely musical effect of any sound depends on its place in what is technically called a 'scale;' the same sound producing absolutely different effects on the ear, according as it is heard in connection with one or another series of sounds.

"It is on this *relative* association of the sounds that all the essentially characteristic effects which are summed up in the phrase 'musical expression,' depend. But why certain associations of sounds have such-and-such effects, is a problem which yet remains to be solved. These effects must indeed, in some way or other, be connected with the well-known arithmetical relations between the rates of vibration of the sounds which form a musical scale. And it is possible—but this is merely a suggestion—that the greater or less mechanical facility with which the vibrating apparatus of the human larynx passes from one state of vibration to another, may have been a primary cause of the greater or less pleasure produced by various sequences of sounds."

But leaving aside these complex questions and confining ourselves to the simpler sounds, we can, at least, see some reasons for the association of certain kinds of sounds with certain states of mind. A scream, for instance, uttered by a young animal, or by one of the members of a community, as a call for assistance, will naturally be loud, prolonged, and high, so as to penetrate to a distance. For Helmholtz has shown [7] that, owing to the shape of the internal cavity of the human ear and its consequent power of resonance, high notes produce a particularly strong impression. When male animals utter sounds in order to please the females, they would naturally employ those which are sweet to the ears of the species; and it appears that the same sounds are often pleasing to widely different animals, owing to the similarity of their nervous systems, as we ourselves perceive in the singing of birds and even in the chirping of certain tree-frogs giving us pleasure. On the other hand, sounds produced in order to strike terror into an enemy, would naturally be harsh or displeasing.

Whether the principle of antithesis has come into play with sounds, as might perhaps have been expected, is doubtful. The interrupted, laughing or tittering sounds made by man and by various kinds of monkeys when pleased, are as different as possible from the prolonged screams of these animals when distressed. The deep grunt of satisfaction uttered by a pig, when pleased with its food, is widely different from its harsh scream of pain or terror. But with the dog, as lately remarked, the bark of anger and that of joy are sounds which by no means stand in opposition to each other; and so it is in some other cases.

There is another obscure point, namely, whether the sounds which are produced under various states of the mind determine the shape of the mouth, or whether its shape is not determined by independent causes, and the sound thus modified. When young infants cry they open their mouths widely, and this, no doubt, is neces-

---

[7] 'Théorie Physiologique de la Musique,' Paris, 1868, p. 146. Helmholtz has also fully discussed in this profound work the relation of the form of the cavity of the mouth to the production of vowel-sounds.

sary for pouring forth a full volume of sound; but the mouth then assumes, from a quite distinct cause, an almost quadrangular shape, depending, as will hereafter be explained, on the firm closing of the eyelids, and consequent drawing up of the upper lip. How far this square shape of the mouth modifies the wailing or crying sound, I am not prepared to say; but we know from the researches of Helmholtz and others that the form of the cavity of the mouth and lips determines the nature and pitch of the vowel sounds which are produced.

It will also be shown in a future chapter that, under the feeling of contempt or disgust, there is a tendency, from intelligible causes, to blow out of the mouth or nostrils, and this produces sounds like *pooh* or *pish*. When any one is startled or suddenly astonished, there is an instantaneous tendency, likewise from an intelligible cause, namely, to be ready for prolonged exertion, to open the mouth widely, so as to draw a deep and rapid inspiration. When the next full expiration follows, the mouth is slightly closed, and the lips, from causes hereafter to be discussed, are somewhat protruded; and this form of the mouth, if the voice be at all exerted, produces, according to Helmholtz, the sound of the vowel *O*. Certainly a deep sound of a prolonged *Oh!* may be heard from a whole crowd of people immediately after witnessing any astonishing spectacle. If, together with surprise, pain be felt, there is a tendency to contract all the muscles of the body, including those of the face, and the lips will then be drawn back; and this will perhaps account for the sound becoming higher and assuming the character of *Ah!* or *Ach!* As fear causes all the muscles of the body to tremble, the voice naturally becomes tremulous, and at the same time husky from the dryness of the mouth, owing to the salivary glands failing to act. Why the laughter of man and the tittering of monkeys should be a rapidly reiterated sound, cannot be explained. During the utterance of these sounds, the mouth is transversely elongated by the corners being drawn backwards and upwards; and of this fact an explanation will be attempted in a future chapter. But the whole subject of the differences of the sounds produced under different states of the mind is so obscure, that I have succeeded in throwing hardly any light on it; and the remarks which I have made, have but little significance.

All the sounds hitherto noticed depend on the respiratory organs; but sounds produced by wholly different means are likewise expressive. Rabbits stamp loudly on the ground as a signal to their comrades; and if a man knows how to do so properly, he may on a quiet evening hear the rabbits answering him all around. These animals, as well as some others, also stamp on the ground when made angry. Porcupines rattle their quills and vibrate their tails when angered; and one behaved in this manner when a live snake was placed in its compartment. The quills on the tail are very different from those on the body: they are short, hollow, thin like a goose-quill, with their ends transversely truncated, so that they are open; they are supported on long, thin, elastic footstalks. Now, when the tail is rapidly shaken, these hollow quills strike against each other and produce, as I heard in the presence of Mr. Bartlett, a peculiar

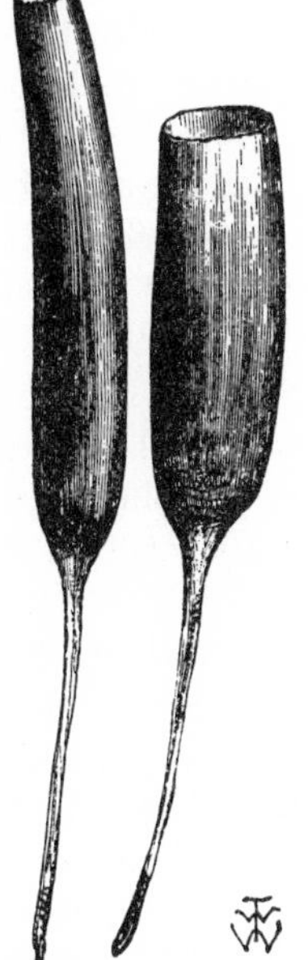

FIG. 11. — Sound-producing quills from the tail of the Porcupine.

continuous sound. We can, I think, understand why porcupines have been provided, through the modification of their protective spines, with this special sound-producing instrument. They are nocturnal animals, and if they scented or heard a prowling beast of prey, it would be a great advantage to them in the dark to give warning to their enemy what they were, and that they were furnished with dangerous spines. They would thus escape being attacked. They are, as I may add, so fully conscious of the power of their weapons, that when enraged they will charge backwards with their spines erected, yet still inclined backwards.

Many birds during their courtship produce diversified sounds by means of specially adapted feathers. Storks, when excited, make a loud clattering noise with their beaks. Some snakes produce a grating or rattling noise. Many insects stridulate by rubbing together specially modified parts of their hard integuments. This stridulation generally serves as a sexual charm or call; but it is likewise used to express different emotions.[8] Every one who has attended to bees knows that their humming changes when they are angry; and this serves as a warning that there is danger of being stung. I have made these few remarks because some writers have laid so much stress on the vocal and respiratory organs as having been specially adapted for expression, that it was advisable to show that sounds otherwise produced serve equally well for the same purpose.

Erection of the dermal appendages.—Hardly any expressive movement is so general as the involuntary erection of the hairs, feathers and other dermal appendages; for it is common throughout three of the great verte-

[8] I have given some details on this subject in my 'Descent of Man,' vol. i. pp. 352, 384.

brate classes. These appendages are erected under the excitement of anger or terror; more especially when these emotions are combined, or quickly succeed each other. The action serves to make the animal appear larger and more frightful to its enemies or rivals, and is generally accompanied by various voluntary movements adapted for the same purpose, and by the utterance of savage sounds. Mr. Bartlett, who has had such wide experience with animals of all kinds, does not doubt that this is the case; but it is a different question whether the power of erection was primarily acquired for this special purpose.

I will first give a considerable body of facts showing how general this action is with mammals, birds and reptiles; retaining what I have to say in regard to man for a future chapter. Mr. Sutton, the intelligent keeper in the Zoological Gardens, carefully observed for me the Chimpanzee and Orang; and he states that when they are suddenly frightened, as by a thunderstorm, or when they are made angry, as by being teased, their hair becomes erect. I saw a chimpanzee who was alarmed at the sight of a black coalheaver, and the hair rose all over his body; he made little starts forward as if to attack the man, without any real intention of doing so, but with the hope, as the keeper remarked, of frightening him. The Gorilla, when enraged, is described by Mr. Ford[9] as having his crest of hair "erect and projecting forward, his nostrils dilated, and his under lip thrown down; at the same time uttering his characteristic yell, designed, it would seem, to terrify his antagonists." I saw the hair on the Anubis baboon, when angered bristling along the back, from the neck to the loins, but not

[9] As quoted in Huxley's 'Evidence as to Man's Place in Nature,' 1863, p. 52.

on the rump or other parts of the body. I took a stuffed snake into the monkey-house, and the hair on several of the species instantly became erect; especially on their tails, as I particularly noticed with the *Cercopithecus nictitans*. Brehm states [10] that the *Midas œdipus* (belonging to the American division) when excited erects its mane, in order, as he adds, to make itself as frightful as possible.

With the Carnivora the erection of the hair seems to be almost universal, often accompanied by threatening movements, the uncovering of the teeth and the utterance of savage growls. In the Herpestes, I have seen the hair on end over nearly the whole body, including the tail; and the dorsal crest is erected in a conspicuous manner by the Hyæna and Proteles. The enraged lion erects his mane. The bristling of the hair along the neck and back of the dog, and over the whole body of the cat, especially on the tail, is familiar to every one. With the cat it apparently occurs only under fear; with the dog, under anger and fear; but not, as far as I have observed, under abject fear, as when a dog is going to be flogged by a severe gamekeeper. If, however, the dog shows fight, as sometimes happens, up goes his hair. I have often noticed that the hair of a dog is particularly liable to rise, if he is half angry and half afraid, as on beholding some object only indistinctly seen in the dusk.

I have been assured by a veterinary surgeon that he has often seen the hair erected on horses and cattle, on which he had operated and was again going to operate. When I showed a stuffed snake to a Peccary, the hair rose in a wonderful manner along its back; and so it does with the boar when enraged. An Elk which gored

---

[10] Illust. Thierleben, 1864, B. i. s. 130.

a man to death in the United States, is described as first brandishing his antlers, squealing with rage and stamping on the ground; "at length his hair was seen to rise and stand on end," and then he plunged forward to the attack.[11] The hair likewise becomes erect on goats, and, as I hear from Mr. Blyth, on some Indian antelopes. I have seen it erected on the hairy Ant-eater; and on the Agouti, one of the Rodents. A female Bat,[12] which reared her young under confinement, when any one looked into the cage "erected the fur on her back, and bit viciously at intruding fingers."

Birds belonging to all the chief Orders ruffle their feathers when angry or frightened. Every one must have seen two cocks, even quite young birds, preparing to fight with erected neck-hackles; nor can these feathers when erected serve as a means of defence, for cock-fighters have found by experience that it is advantageous to trim them. The male Ruff (*Machetes pugnax*) likewise erects its collar of feathers when fighting. When a dog approaches a common hen with her chickens, she spreads out her wings, raises her tail, ruffles all her feathers, and looking as ferocious as possible, dashes at the intruder. The tail is not always held in exactly the same position; it is sometimes so much erected, that the central feathers, as in the accompanying drawing, almost touch the back. Swans, when angered, likewise raise their wings and tail, and erect their feathers. They open their beaks, and make by paddling little rapid starts forwards, against any one who approaches the water's edge too closely. Tropic birds [13] when disturbed on their nests

---

[11] The Hon. J. Caton, Ottawa Acad. of Nat. Sciences, May, 1868, pp. 36, 40. For the *Capra Ægagrus*, 'Land and Water,' 1867, p. 37.

[12] 'Land and Water,' July 20, 1867, p. 659.

[13] *Phaeton rubricauda:* 'Ibis,' vol. iii. 1861, p. 180.

are said not to fly away, but "merely to stick out their feathers and scream." The Barn-owl, when approached "instantly swells out its plumage, extends its wings and tail, hisses and clacks its mandibles with force and rapidity."[14] So do other kinds of owls. Hawks, as I am

FIG. 12.—Hen driving away a dog from her chickens. Drawn from life by Mr. Wood.

informed by Mr. Jenner Weir, likewise ruffle their feathers, and spread out their wings and tail under similar circumstances. Some kinds of parrots erect their feathers; and I have seen this action in the Cassowary, when angered at the sight of an Ant-eater. Young cuckoos in the nest, raise their feathers, open their mouths widely, and make themselves as frightful as possible.

---

[14] On the *Strix flammea*, Audubon, 'Ornithological Biography,' 1864, vol. ii. p. 407. I have observed other cases in the Zoological Gardens.

Small birds, also, as I hear from Mr. Weir, such as various finches, buntings and warblers, when angry,

FIG. 13.—Swan driving away an intruder. Drawn from life by Mr. Wood.

ruffle all their feathers, or only those round the neck; or they spread out their wings and tail-feathers. With their plumage in this state, they rush at each other with open beaks and threatening gestures. Mr. Weir concludes from his large experience that the erection of the feathers is caused much more by anger than by fear. He gives as an instance a hybrid goldfinch of a most irascible disposition, which when approached too closely by a servant, instantly assumes the appearance of a ball of ruffled feathers. He believes that birds when frightened, as a general rule, closely adpress all their feathers, and their consequently diminished size is often astonishing.

As soon as they recover from their fear or surprise, the first thing which they do is to shake out their feathers. The best instances of this adpression of the feathers and apparent shrinking of the body from fear, which Mr. Weir has noticed, has been in the quail and grass-parrakeet.[15] The habit is intelligible in these birds from their being accustomed, when in danger, either to squat on the ground or to sit motionless on a branch, so as to escape detection. Though, with birds, anger may be the chief and commonest cause of the erection of the feathers, it is probable that young cuckoos when looked at in the nest, and a hen with her chickens when approached by a dog, feel at least some terror. Mr. Tegetmeier informs me that with game-cocks, the erection of the feathers on the head has long been recognized in the cock-pit as a sign of cowardice.

The males of some lizards, when fighting together during their courtship, expand their throat pouches or frills, and erect their dorsal crests.[16] But Dr. Günther does not believe that they can erect their separate spines or scales.

We thus see how generally throughout the two higher vertebrate classes, and with some reptiles, the dermal appendages are erected under the influence of anger and fear. The movement is effected, as we know from Kölliker's interesting discovery, by the contraction of minute, unstriped, involuntary muscles,[17] often called *arrectores pili*, which are attached to the capsules of the separate

---

[15] *Melopsittacus undulatus.* See an account of its habits by Gould, 'Handbook of Birds of Australia,' 1865, vol. ii. p. 82.

[16] See, for instance, the account which I have given ('Descent of Man,' vol. ii. p. 32) of an Anolis and Draco.

[17] These muscles are described in his well-known works. I am greatly indebted to this distinguished observer for having given me in a letter information on this same subject.

hairs, feathers, &c. By the contraction of these muscles the hairs can be instantly erected, as we see in a dog, being at the same time drawn a little out of their sockets; they are afterwards quickly depressed. The vast number of these minute muscles over the whole body of a hairy quadruped is astonishing. The erection of the hair is, however, aided in some cases, as with that on the head of a man, by the striped and voluntary muscles of the underlying *panniculus carnosus*. It is by the action of these latter muscles, that the hedgehog erects its spines. It appears, also, from the researches of Leydig[18] and others, that striped fibres extend from the panniculus to some of the larger hairs, such as the vibrissæ of certain quadrupeds. The *arrectores pili* contract not only under the above emotions, but from the application of cold to the surface. I remember that my mules and dogs, brought from a lower and warmer country, after spending a night on the bleak Cordillera, had the hair all over their bodies as erect as under the greatest terror. We see the same action in our own *goose-skin* during the chill before a fever-fit. Mr. Lister has also found,[19] that tickling a neighbouring part of the skin causes the erection and protrusion of the hairs.

From these facts it is manifest that the erection of the dermal appendages is a reflex action, independent of the will; and this action must be looked at, when, occurring under the influence of anger or fear, not as a power acquired for the sake of some advantage, but as an incidental result, at least to a large extent, of the sensorium being affected. The result, in as far as it is

[18] 'Lehrbuch der Histologie des Menschen,' 1857, s. 82. I owe to Prof. W. Turner's kindness an extract from this work.

[19] 'Quarterly Journal of Microscopical Science,' 1853, vol. i. p. 262.

incidental, may be compared with the profuse sweating from an agony of pain or terror. Nevertheless, it is remarkable how slight an excitement often suffices to cause the hair to become erect; as when two dogs pretend to fight together in play. We have, also, seen in a large number of animals, belonging to widely distinct classes, that the erection of the hair or feathers is almost always accompanied by various voluntary movements—by threatening gestures, opening the mouth, uncovering the teeth, spreading out of the wings and tail by birds, and by the utterance of harsh sounds; and the purpose of these voluntary movements is unmistakable. Therefore it seems hardly credible that the co-ordinated erection of the dermal appendages, by which the animal is made to appear larger and more terrible to its enemies or rivals, should be altogether an incidental and purposeless result of the disturbance of the sensorium. This seems almost as incredible as that the erection by the hedgehog of its spines, or of the quills by the porcupine, or of the ornamental plumes by many birds during their courtship, should all be purposeless actions.

We here encounter a great difficulty. How can the contraction of the unstriped and involuntary *arrectores pili* have been co-ordinated with that of various voluntary muscles for the same special purpose? If we could believe that the arrectores primordially had been voluntary muscles, and had since lost their stripes and become involuntary, the case would be comparatively simple. I am not, however, aware that there is any evidence in favour of this view; although the reversed transition would not have presented any great difficulty, as the voluntary muscles are in an unstriped condition in the embryos of the higher animals, and in the larvæ of some crustaceans. Moreover in the deeper layers of the skin of adult birds, the muscular network is, according to

Leydig,[20] in a transitional condition; the fibres exhibiting only indications of transverse striation.

Another explanation seems possible. We may admit that originally the *arrectores pili* were slightly acted on in a direct manner, under the influence of rage and terror, by the disturbance of the nervous system; as is undoubtedly the case with our so-called *goose-skin* before a fever-fit. Animals have been repeatedly excited by rage and terror during many generations; and consequently the direct effects of the disturbed nervous system on the dermal appendages will almost certainly have been increased through habit and through the tendency of nerve-force to pass readily along accustomed channels. We shall find this view of the force of habit strikingly confirmed in a future chapter, where it will be shown that the hair of the insane is affected in an extraordinary manner, owing to their repeated accesses of fury and terror. As soon as with animals the power of erection had thus been strengthened or increased, they must often have seen the hairs or feathers erected in rival and enraged males, and the bulk of their bodies thus increased. In this case it appears possible that they might have wished to make themselves appear larger and more terrible to their enemies, by voluntarily assuming a threatening attitude and uttering harsh cries; such attitudes and utterances after a time becoming through habit instinctive. In this manner actions performed by the contraction of voluntary muscles might have been combined for the same special purpose with those effected by involuntary muscles. It is even possible that animals, when excited and dimly conscious of some change in the state of their hair, might act on it by repeated exertions of their attention and will; for we have

[20] 'Lehrbuch der Histologie,' 1857, s. 82.

reason to believe that the will is able to influence in an obscure manner the action of some unstriped or involuntary muscles, as in the period of the peristaltic movements of the intestines, and in the contraction of the bladder. Nor must we overlook the part which variation and natural selection may have played; for the males which succeeded in making themselves appear the most terrible to their rivals, or to their other enemies, if not of overwhelming power, will on an average have left more offspring to inherit their characteristic qualities, whatever these may be and however first acquired, than have other males.

The inflation of the body, and other means of exciting fear in an enemy.—Certain Amphibians and Reptiles, which either have no spines to erect, or no muscles by which they can be erected, enlarge themselves when alarmed or angry by inhaling air. This is well known to be the case with toads and frogs. The latter animal is made, in Æsop's fable of the 'Ox and the Frog,' to blow itself up from vanity and envy until it burst. This action must have been observed during the most ancient times, as, according to Mr. Hensleigh Wedgwood,[21] the word *toad* expresses in all the languages of Europe the habit of swelling. It has been observed with some of the exotic species in the Zoological Gardens; and Dr. Günther believes that it is general throughout the group. Judging from analogy, the primary purpose probably was to make the body appear as large and frightful as possible to an enemy; but another, and perhaps more important secondary advantage is thus gained. When frogs are seized by snakes, which are their chief enemies, they enlarge themselves wonderfully; so that if the snake be of small size, as Dr. Günther informs me, it cannot swallow the frog, which thus escapes being devoured.

---

[21] 'Dictionary of English Etymology,' p. 403.

Chameleons and some other lizards inflate themselves when angry. Thus a species inhabiting Oregon, the *Tapaya Douglasii*, is slow in its movements and does not bite, but has a ferocious aspect; "when irritated it springs in a most threatening manner at anything pointed at it, at the same time opening its mouth wide and hissing audibly, after which it inflates its body, and shows other marks of anger." [22]

Several kinds of snakes likewise inflate themselves when irritated. The puff-adder (*Clotho arietans*) is remarkable in this respect; but I believe, after carefully watching these animals, that they do not act thus for the sake of increasing their apparent bulk, but simply for inhaling a large supply of air, so as to produce their surprisingly loud, harsh, and prolonged hissing sound. The Cobras-de-capello, when irritated, enlarge themselves a little, and hiss moderately; but, at the same time they lift their heads aloft, and dilate by means of their elongated anterior ribs, the skin on each side of the neck into a large flat disk,—the so-called hood. With their widely opened mouths, they then assume a terrific aspect. The benefit thus derived ought to be considerable, in order to compensate for the somewhat lessened rapidity (though this is still great) with which, when dilated, they can strike at their enemies or prey; on the same principle that a broad, thin piece of wood cannot be moved through the air so quickly as a small round stick. An innocuous snake, the *Tropidonotus macrophthalmus*, an inhabitant of India, likewise dilates its neck when irritated; and consequently is often mistaken for its compatriot, the deadly Cobra.[23] This resemblance perhaps serves as some protection to the Tropidonotus.

---

[22] See the account of the habits of this animal by Dr. Cooper, as quoted in 'Nature,' April 27, 1871, p. 512.

[23] Dr. Günther, 'Reptiles of British India,' p. 262.

Another innocuous species, the Dasypeltis of South Africa, blows itself out, distends its neck, hisses and darts at an intruder.[24] Many other snakes hiss under similar circumstances. They also rapidly vibrate their protruded tongues; and this may aid in increasing their terrific appearance.

Snakes possess other means of producing sounds besides hissing. Many years ago I observed in South America that a venomous Trigonocephalus, when disturbed, rapidly vibrated the end of its tail, which striking against the dry grass and twigs produced a rattling noise that could be distinctly heard at the distance of six feet.[25] The deadly and fierce *Echis carinata* of India produces "a curious prolonged, almost hissing sound" in a very different manner, namely by rubbing "the sides of the folds of its body against each other," whilst the head remains in almost the same position. The scales on the sides, and not on other parts of the body, are strongly keeled, with the keels toothed like a saw; and as the coiled-up animal rubs its sides together, these grate against each other.[26] Lastly, we have the well-known case of the Rattle-snake. He who has merely shaken the rattle of a dead snake, can form no just idea of the sound produced by the living animal. Professor Shaler states that it is indistinguishable from that made by the male of a large Cicada (an Homopterous insect), which inhabits the same district.[27] In the Zoological

[24] Mr. J. Mansel Weale, 'Nature,' April 27, 1871, p. 508.

[25] 'Journal of Researches during the Voyage of the "Beagle,"' 1845, p. 96. I have compared the rattling thus produced with that of the Rattle-snake.

[26] See the account by Dr. Anderson, Proc. Zool. Soc. 1871, p. 196.

[27] The 'American Naturalist,' Jan. 1872, p. 32. I regret that I cannot follow Prof. Shaler in believing that the rattle has been developed, by the aid of natural selection, for the sake of producing sounds which deceive and attract birds, so that they may serve as prey to the snake.

Gardens, when the rattle-snakes and puff-adders were greatly excited at the same time, I was much struck at the similarity of the sound produced by them; and although that made by the rattle-snake is louder and shriller than the hissing of the puff-adder, yet when standing at some yards distance I could scarcely distinguish the two. For whatever purpose the sound is produced by the one species, I can hardly doubt that it serves for the same purpose in the other species; and I conclude from the threatening gestures made at the same time by many snakes, that their hissing,—the rattling of the rattle-snake and of the tail of the Trigonocephalus,—the grating of the scales of the Echis,—and the dilatation of the hood of the Cobra,—all subserve the same end, namely, to make them appear terrible to their enemies.[28]

It seems at first a probable conclusion that venomous snakes, such as the foregoing, from being already so well defended by their poison-fangs, would never be attacked by any enemy; and consequently would have

---

I do not, however, wish to doubt that the sounds may occasionally subserve this end. But the conclusion at which I have arrived, viz. that the rattling serves as a warning to would-be devourers, appears to me much more probable, as it connects together various classes of facts. If this snake had acquired its rattle and the habit of rattling, for the sake of attracting prey, it does not seem probable that it would have invariably used its instrument when angered or disturbed. Prof. Shaler takes nearly the same view as I do of the manner of development of the rattle; and I have always held this opinion since observing the Trigonocephalus in South America.

[28] From the accounts lately collected, and given in the 'Journal of the Linnean Society,' by Mrs. Barber, on the habits of the snakes of South Africa; and from the accounts published by several writers, for instance by Lawson, of the rattle-snake in North America,—it does not seem improbable that the terrific appearance of snakes and the sounds produced by them, may likewise serve in procuring prey, by paralysing, or as it is sometimes called fascinating, the smaller animals.

no need to excite additional terror. But this is far from being the case, for they are largely preyed on in all quarters of the world by many animals. It is well known that pigs are employed in the United States to clear districts infested with rattle-snakes, which they do most effectually.[29] In England the hedgehog attacks and devours the viper. In India, as I hear from Dr. Jerdon, several kinds of hawks, and at least one mammal, the Herpestes, kill cobras and other venomous species;[30] and so it is in South Africa. Therefore it is by no means improbable that any sounds or signs by which the venomous species could instantly make themselves recognized as dangerous, would be of more service to them than to the innocuous species which would not be able, if attacked, to inflict any real injury.

Having said thus much about snakes, I am tempted to add a few remarks on the means by which the rattle of the rattle-snake was probably developed. Various animals, including some lizards, either curl or vibrate their tails when excited. This is the case with many kinds of snakes.[31] In the Zoological Gardens, an in-

[29] See the account by Dr. R. Brown, in Proc. Zool. Soc. 1871, p. 39. He says that as soon as a pig sees a snake it rushes upon it; and a snake makes off immediately on the appearance of a pig.

[30] Dr. Günther remarks ('Reptiles of British India,' p. 340) on the destruction of cobras by the ichneumon or herpestes, and whilst the cobras are young by the jungle-fowl. It is well known that the peacock also eagerly kills snakes.

[31] Prof. Cope enumerates a number of kinds in his 'Method of Creation of Organic Types,' read before the American Phil. Soc., December 15th, 1871, p. 20. Prof. Cope takes the same view as I do of the use of the gestures and sounds made by snakes. I briefly alluded to this subject in the last edition of my 'Origin of Species.' Since the passages in the text above have been printed, I have been pleased to find that Mr. Henderson ('The American Naturalist,' May, 1872, p. 260) also takes a similar view of the use of the rattle, namely "in preventing an attack from being made."

nocuous species, the *Coronella Sayi*, vibrates its tail so rapidly that it becomes almost invisible. The Trigonocephalus, before alluded to, has the same habit; and the extremity of its tail is a little enlarged, or ends in a bead. In the Lachesis, which is so closely allied to the rattle-snake that it was placed by Linnæus in the same genus, the tail ends in a single, large, lancet-shaped point or scale. With some snakes the skin, as Professor Shaler remarks, "is more imperfectly detached from the region about the tail than at other parts of the body." Now if we suppose that the end of the tail of some ancient American species was enlarged, and was covered by a single large scale, this could hardly have been cast off at the successive moults. In this case it would have been permanently retained, and at each period of growth, as the snake grew larger, a new scale, larger than the last, would have been formed above it, and would likewise have been retained. The foundation for the development of a rattle would thus have been laid; and it would have been habitually used, if the species, like so many others, vibrated its tail whenever it was irritated. That the rattle has since been specially developed to serve as an efficient sound-producing instrument, there can hardly be a doubt; for even the vertebræ included within the extremity of the tail have been altered in shape and cohere. But there is no greater improbability in various structures, such as the rattle of the rattle-snake,—the lateral scales of the Echis,—the neck with the included ribs of the Cobra,—and the whole body of the puff-adder,—having been modified for the sake of warning and frightening away their enemies, than in a bird, namely, the wonderful Secretary-hawk (*Gypogeranus*) having had its whole frame modified for the sake of killing snakes with impunity. It is highly probable, judging from what we have before seen, that this bird would ruffle its feathers whenever it attacked a snake; and it is certain that the Herpestes, when it eagerly rushes to attack a snake, erects the hair all over its body, and especially that on its tail.[32] We have also seen that some porcupines, when angered or alarmed at the sight of a snake, rapidly vibrate their tails, thus producing a peculiar sound by the striking together of the hollow quills. So that here both the attackers and the attacked endeavour to make themselves as dreadful as possible to each other; and both possess for this purpose specialised means, which, oddly enough, are nearly the same in some of these cases. Finally we can see that if, on the one hand, those individual snakes, which were best able to frighten away their enemies, escaped best from being devoured; and if, on the other hand, those individuals of the attacking enemy survived in larger numbers which were the best fitted for the dangerous task of killing and devouring venomous snakes;—then in the one case as in the other, beneficial variations, supposing the characters in question to vary, would commonly have been preserved through the survival of the fittest.

The Drawing back and pressure of the Ears to the Head.—The ears through their movements are highly expressive in many animals; but in some, such as man, the higher apes, and many ruminants, they fail in this respect. A slight difference in position serves to express in the plainest manner a different state of mind, as we may daily see in the dog; but we are here concerned only with the ears being drawn closely backwards and pressed to the head. A savage frame of mind is thus shown, but only in the case of those animals which fight

---

[32] Mr. des Vœux, in Proc. Zool. Soc. 1871, p. 3.

with their teeth; and the care which they take to prevent their ears being seized by their antagonists, accounts for this position. Consequently, through habit and association, whenever they feel slightly savage, or pretend in their play to be savage, their ears are drawn back. That this is the true explanation may be inferred from the relation which exists in very many animals between their manner of fighting and the retraction of their ears.

All the Carnivora fight with their canine teeth, and all, as far as I have observed, draw their ears back when feeling savage. This may be continually seen with dogs when fighting in earnest, and with puppies fighting in play. The movement is different from the falling down and slight drawing back of the ears, when a dog feels pleased and is caressed by his master. The retraction of the ears may likewise be seen in kittens fighting together in their play, and in full-grown cats when really savage, as before illustrated in fig. 9 (p. 58). Although their ears are thus to a large extent protected, yet they often get much torn in old male cats during their mutual battles. The same movement is very striking in tigers, leopards, &c., whilst growling over their food in menageries. The lynx has remarkably long ears; and their retraction, when one of these animals is approached in its cage, is very conspicuous, and is eminently expressive of its savage disposition. Even one of the Eared Seals, the *Otaria pusilla*, which has very small ears, draws them backwards, when it makes a savage rush at the legs of its keeper.

When horses fight together they use their incisors for biting, and their fore-legs for striking, much more than they do their hind-legs for kicking backwards. This has been observed when stallions have broken loose and have fought together, and may likewise be inferred from the kind of wounds which they inflict on each other.

Every one recognizes the vicious appearance which the drawing back of the ears gives to a horse. This movement is very different from that of listening to a sound behind. If an ill-tempered horse in a stall is inclined to kick backwards, his ears are retracted from habit, though he has no intention or power to bite. But when a horse throws up both hind-legs in play, as when entering an open field, or when just touched by the whip, he does not generally depress his ears, for he does not then feel vicious. Guanacoes fight savagely with their teeth; and they must do so frequently, for I found the hides of several which I shot in Patagonia deeply scored. So do camels; and both these animals, when savage, draw their ears closely backwards. Guanacoes, as I have noticed, when not intending to bite, but merely to spit their offensive saliva from a distance at an intruder, retract their ears. Even the hippopotamus, when threatening with its widely-open enormous mouth a comrade, draws back its small ears, just like a horse.

Now what a contrast is presented between the foregoing animals and cattle, sheep, or goats, which never use their teeth in fighting, and never draw back their ears when enraged! Although sheep and goats appear such placid animals, the males often join in furious contests. As deer form a closely related family, and as I did not know that they ever fought with their teeth, I was much surprised at the account given by Major Ross King of the Moose-deer in Canada. He says, when "two males chance to meet, laying back their ears and gnashing their teeth together, they rush at each other with appalling fury." [33] But Mr. Bartlett informs me that some species of deer fight savagely with their teeth,

[33] 'The Sportsman and Naturalist in Canada,' 1866, p. 53.
p. 53.

so that the drawing back of the ears by the moose accords with our rule. Several kinds of kangaroos, kept in the Zoological Gardens, fight by scratching with their fore-feet and by kicking with their hind-legs; but they never bite each other, and the keepers have never seen them draw back their ears when angered. Rabbits fight chiefly by kicking and scratching, but they likewise bite each other; and I have known one to bite off half the tail of its antagonist. At the commencement of their battles they lay back their ears, but afterwards, as they bound over and kick each other, they keep their ears erect, or move them much about.

Mr. Bartlett watched a wild boar quarrelling rather savagely with his sow; and both had their mouths open and their ears drawn backwards. But this does not appear to be a common action with domestic pigs when quarrelling. Boars fight together by striking upwards with their tusks; and Mr. Bartlett doubts whether they then draw back their ears. Elephants, which in like manner fight with their tusks, do not retract their ears, but, on the contrary, erect them when rushing at each other or at an enemy.

The rhinoceroses in the Zoological Gardens fight with their nasal horns, and have never been seen to attempt biting each other except in play; and the keepers are convinced that they do not draw back their ears, like horses and dogs, when feeling savage. The following statement, therefore, by Sir S. Baker[34] is inexplicable, namely, that a rhinoceros, which he shot in North Africa, "had no ears; they had been bitten off close to the head by another of the same species while fighting; and this mutilation is by no means uncommon."

Lastly, with respect to monkeys. Some kinds, which

[34] 'The Nile Tributaries of Abyssinia,' 1867, p. 443.

have moveable ears, and which fight with their teeth—for instance the *Cercopithecus ruber*—draw back their ears when irritated just like dogs; and they then have a very spiteful appearance. Other kinds, as the *Inuus ecaudatus*, apparently do not thus act. Again, other kinds—and this is a great anomaly in comparison with most other animals—retract their ears, show their teeth, and jabber, when they are pleased by being caressed. I observed this in two or three species of Macacus, and in the *Cynopithecus niger*. This expression, owing to our familiarity with dogs, would never be recognized as one of joy or pleasure by those unacquainted with monkeys.

Erection of the Ears.—This movement requires hardly any notice. All animals which have the power of freely moving their ears, when they are startled, or when they closely observe any object, direct their ears to the point towards which they are looking, in order to hear any sound from this quarter. At the same time they generally raise their heads, as all their organs of sense are there situated, and some of the smaller animals rise on their hind-legs. Even those kinds which squat on the ground or instantly flee away to avoid danger, generally act momentarily in this manner, in order to ascertain the source and nature of the danger. The head being raised, with erected ears and eyes directed forwards, gives an unmistakable expression of close attention to any animal.

CHAPTER V.

SPECIAL EXPRESSIONS OF ANIMALS.

The Dog, various expressive movements of—Cats—Horses—Ruminants—Monkeys, their expression of joy and affection—Of pain—Anger—Astonishment and Terror.

The Dog.—I have already described (figs. 5 and 7) the appearance of a dog approaching another dog with hostile intentions, namely, with erected ears, eyes intently directed forwards, hair on the neck and back bristling, gait remarkably stiff, with the tail upright and rigid. So familiar is this appearance to us, that an angry man is sometimes said "to have his back up." Of the above points, the stiff gait and upright tail alone require further discussion. Sir C. Bell remarks [1] that, when a tiger or wolf is struck by its keeper and is suddenly roused to ferocity, "every muscle is in tension, and the limbs are in an attitude of strained exertion, prepared to spring." This tension of the muscles and consequent stiff gait may be accounted for on the principle of associated habit, for anger has continually led to fierce struggles, and consequently to all the muscles of the body having been violently exerted. There is also reason to suspect that the muscular system requires some short preparation, or some degree of innervation, before being brought into strong action. My own sensations

[1] 'The Anatomy of Expression,' 1844, p. 190.

lead me to this inference; but I cannot discover that it is a conclusion admitted by physiologists. Sir J. Paget, however, informs me that when muscles are suddenly contracted with the greatest force, without any preparation, they are liable to be ruptured, as when a man slips unexpectedly; but that this rarely occurs when an action, however violent, is deliberately performed.

With respect to the upright position of the tail, it seems to depend (but whether this is really the case I know not) on the elevator muscles being more powerful than the depressors, so that when all the muscles of the hinder part of the body are in a state of tension, the tail is raised. A dog in cheerful spirits, and trotting before his master with high, elastic steps, generally carries his tail aloft, though it is not held nearly so stiffly as when he is angered. A horse when first turned out into an open field, may be seen to trot with long elastic strides, the head and tail being held high aloft. Even cows when they frisk about from pleasure, throw up their tails in a ridiculous fashion. So it is with various animals in the Zoological Gardens. The position of the tail, however, in certain cases, is determined by special circumstances; thus as soon as a horse breaks into a gallop, at full speed, he always lowers his tail, so that as little resistance as possible may be offered to the air.

When a dog is on the point of springing on his antagonist, he utters a savage growl; the ears are pressed closely backwards, and the upper lip (fig. 14) is retracted out of the way of his teeth, especially of his canines. These movements may be observed with dogs and puppies in their play. But if a dog gets really savage in his play, his expression immediately changes. This, however, is simply due to the lips and ears being drawn back with much greater energy. If a dog only snarls at another, the lip is generally retracted on one side alone, namely towards his enemy.

The movements of a dog whilst exhibiting affection towards his master were described (figs. 6 and 8) in our second chapter. These consist in the head and whole body being lowered and thrown into flexuous movements, with the tail extended and wagged from side to side. The ears fall down and are drawn somewhat backwards, which causes the eyelids to be elongated, and alters the

FIG. 14.—Head of snarling Dog. From life, by Mr. Wood.

whole appearance of the face. The lips hang loosely, and the hair remains smooth. All these movements or gestures are explicable, as I believe, from their standing in complete antithesis to those naturally assumed by a savage dog under a directly opposite state of mind. When a man merely speaks to, or just notices, his dog, we see the last vestige of these movements in a slight wag of the tail, without any other movement of the body, and without even the ears being lowered. Dogs also exhibit their affection by desiring to rub against their masters, and to be rubbed or patted by them.

Gratiolet explains the above gestures of affection in the following manner: and the reader can judge whether the explanation appears satisfactory. Speaking of animals in general, including the dog, he says,[2] "C'est toujours la partie la plus sensible de leurs corps qui recherche les caresses ou les donne. Lorsque toute la longueur des flancs et du corps est sensible, l'animal serpente et rampe sous les caresses; et ces ondulations se propageant le long des muscles analogues des segments jusqu'aux extrémités de la colonne vertébrale, la queue se ploie et s'agite." Further on, he adds, that dogs, when feeling affectionate, lower their ears in order to exclude all sounds, so that their whole attention may be concentrated on the caresses of their master!

Dogs have another and striking way of exhibiting their affection, namely, by licking the hands or faces of their masters. They sometimes lick other dogs, and then it is always their chops. I have also seen dogs licking cats with whom they were friends. This habit probably originated in the females carefully licking their puppies—the dearest object of their love—for the sake of cleansing them. They also often give their puppies, after a short absence, a few cursory licks, apparently from affection. Thus the habit will have become associated with the emotion of love, however it may afterwards be aroused. It is now so firmly inherited or innate, that it is transmitted equally to both sexes. A female terrier of mine lately had her puppies destroyed,

---

[2] 'De la Physionomie,' 1865, pp. 187, 218.

and though at all times a very affectionate creature, I was much struck with the manner in which she then tried to satisfy her instinctive maternal love by expending it on me; and her desire to lick my hands rose to an insatiable passion.

The same principle probably explains why dogs, when feeling affectionate, like rubbing against their masters and being rubbed or patted by them, for from the nursing of their puppies, contact with a beloved object has become firmly associated in their minds with the emotion of love.

The feeling of affection of a dog towards his master is combined with a strong sense of submission, which is akin to fear. Hence dogs not only lower their bodies and crouch a little as they approach their masters, but sometimes throw themselves on the ground with their bellies upwards. This is a movement as completely opposite as is possible to any show of resistance. I formerly possessed a large dog who was not at all afraid to fight with other dogs; but a wolf-like shepherd-dog in the neighbourhood, though not ferocious and not so powerful as my dog, had a strange influence over him. When they met on the road, my dog used to run to meet him, with his tail partly tucked in between his legs and hair not erected; and then he would throw himself on the ground, belly upwards. By this action he seemed to say more plainly than by words, "Behold, I am your slave."

A pleasurable and excited state of mind, associated with affection, is exhibited by some dogs in a very peculiar manner; namely, by grinning. This was noticed long ago by Somerville, who says,

> "And with a courtly grin, the fawning hound
> Salutes thee cow'ring, his wide op'ning nose
> Upward he curls, and his large sloe-back eyes
> Melt in soft blandishments, and humble joy.'
>
> *The Chase*, book i.

Sir W. Scott's famous Scotch greyhound, Maida, had this habit, and it is common with terriers. I have also seen it in a Spitz and in a sheep-dog. Mr. Riviere, who has particularly attended to this expression, informs me that it is rarely displayed in a perfect manner, but is quite common in a lesser degree. The upper lip during the act of grinning is retracted, as in snarling, so that the canines are exposed, and the ears are drawn backwards; but the general appearance of the animal clearly shows that anger is not felt. Sir C. Bell [3] remarks "Dogs, in their expression of fondness, have a slight eversion of the lips, and grin and sniff amidst their gambols, in a way that resembles laughter." Some persons speak of the grin as a smile, but if it had been really a smile, we should see a similar, though more pronounced, movement of the lips and ears, when dogs utter their bark of joy; but this is not the case, although a bark of joy often follows a grin. On the other hand, dogs, when playing with their comrades or masters, almost always pretend to bite each other; and they then retract, though not energetically, their lips and ears. Hence I suspect that there is a tendency in some dogs, whenever they feel lively pleasure combined with affection, to act through habit and association on the same muscles, as in playfully biting each other, or their masters' hands.

I have described, in the second chapter, the gait and appearance of a dog when cheerful, and the marked antithesis presented by the same animal when dejected and disappointed, with his head, ears, body, tail, and chops drooping, and eyes dull. Under the expectation of any great pleasure, dogs bound and jump about in an extravagant manner, and bark for joy. The tendency to bark under this state of mind is inherited, or runs in

[3] 'The Anatomy of Expression,' 1844, p. 140.

the breed: greyhounds rarely bark, whilst the Spitz-dog barks so incessantly on starting for a walk with his master that he becomes a nuisance.

An agony of pain is expressed by dogs in nearly the same way as by many other animals, namely, by howling, writhing, and contortions of the whole body.

Attention is shown by the head being raised, with the ears erected, and eyes intently directed towards the object or quarter under observation. If it be a sound and the source is not known, the head is often turned obliquely from side to side in a most significant manner, apparently in order to judge with more exactness from what point the sound proceeds. But I have seen a dog greatly surprised at a new noise, turning his head to one side through habit, though he clearly perceived the source of the noise. Dogs, as formerly remarked, when their attention is in any way aroused, whilst watching some object, or attending to some sound, often lift up one paw (fig. 4) and keep it doubled up, as if to make a slow and stealthy approach.

A dog under extreme terror will throw himself down, howl, and void his excretions; but the hair, I believe, does not become erect unless some anger is felt. I have seen a dog much terrified at a band of musicians who were playing loudly outside the house, with every muscle of his body trembling, with his heart palpitating so quickly that the beats could hardly be counted, and panting for breath with widely open mouth, in the same manner as a terrified man does. Yet this dog had not exerted himself; he had only wandered slowly and restlessly about the room, and the day was cold.

Even a very slight degree of fear is invariably shown by the tail being tucked in between the legs. This tucking in of the tail is accompanied by the ears being drawn backwards; but they are not pressed closely to the head, as in snarling, and they are not lowered, as when a dog is pleased or affectionate. When two young dogs chase each other in play, the one that runs away always keeps his tail tucked inwards. So it is when a dog, in the highest spirits, careers like a mad creature round and round his master in circles, or in figures of eight. He then acts as if another dog were chasing him. This curious kind of play, which must be familiar to every one who has attended to dogs, is particularly apt to be excited, after the animal has been a little startled or frightened, as by his master suddenly jumping out on him in the dusk. In this case, as well as when two young dogs are chasing each other in play, it appears as if the one that runs away was afraid of the other catching him by the tail; but as far as I can find out, dogs very rarely catch each other in this manner. I asked a gentleman, who had kept foxhounds all his life, and he applied to other experienced sportsmen, whether they had ever seen hounds thus seize a fox; but they never had. It appears that when a dog is chased, or when in danger of being struck behind, or of anything falling on him, in all these cases he wishes to withdraw as quickly as possible his whole hind-quarters, and that from some sympathy or connection between the muscles, the tail is then drawn closely inwards.

A similarly connected movement between the hind-quarters and the tail may be observed in the hyæna. Mr. Bartlett informs me that when two of these animals fight together, they are mutually conscious of the wonderful power of each other's jaws, and are extremely cautious. They well know that if one of their legs were seized, the bone would instantly be crushed into atoms; hence they approach each other kneeling, with their legs turned as much as possible inwards, and with their whole bodies bowed, so as not to present any salient point; the tail at the same time being closely tucked in between the legs. In this attitude they approach each other side-ways, or even partly backwards. So again with deer, several of the species, when savage and fighting, tuck in their tails. When one horse in a field tries to bite the hind-quarters of another in play, or when a rough boy strikes a donkey from behind, the hind-quarters and the tail are drawn in, though it does not appear as if this were done merely to save the tail from being injured. We have also seen the reverse of these movements; for when an animal trots with high elastic steps, the tail is almost always carried aloft.

As I have said, when a dog is chased and runs away, he keeps his ears directed backwards but still open; and this is clearly done for the sake of hearing the footsteps of his pursuer. From habit the ears are often held in this same position, and the tail tucked in, when the danger is obviously in front. I have repeatedly noticed, with a timid terrier of mine, that when she is afraid of some object in front, the nature of which she perfectly knows and does not need to reconnoitre, yet she will for a long time hold her ears and tail in this position, looking the image of discomfort. Discomfort, without any fear, is similarly expressed: thus, one day I went out of doors, just at the time when this same dog knew that her dinner would be brought. I did not call her, but she wished much to accompany me, and at the same time she wished much for her dinner; and there she stood, first looking one way and then the other, with her tail tucked in and ears drawn back, presenting an unmistakable appearance of perplexed discomfort.

Almost all the expressive movements now described, with the exception of the grinning from joy, are innate or instinctive, for they are common to all the individuals, young and old, of all the breeds. Most of them are likewise common to the aboriginal parents of the dog, namely the wolf and jackal; and some of them to other species of the same group. Tamed wolves and jackals, when caressed by their masters, jump about for joy, wag their tails, lower their ears, lick their master's hands, crouch down, and even throw themselves on the ground belly upwards.[4] I have seen a rather fox-like African jackal, from the Gaboon, depress its ears when caressed. Wolves and jackals, when frightened, certainly tuck in their tails; and a tamed jackal has been described as careering round his master in circles and figures of eight, like a dog, with his tail between his legs.

It has been stated [5] that foxes, however tame, never display any of the above expressive movements; but this is not strictly accurate. Many years ago I observed in the Zoological Gardens, and recorded the fact at the time, that a very tame English fox, when caressed by the keeper, wagged its tail, depressed its ears, and then threw itself on the ground, belly upwards. The black fox of North America likewise depressed its ears in a slight degree. But I believe that foxes never lick the hands of their masters, and I have been assured that when frightened they never tuck in their tails. If the explanation which I have given of the expression of affection in dogs be admitted, then it would appear that animals which have never been domesticated—namely wolves, jackals, and even foxes—have nevertheless ac-

---

[4] Many particulars are given by Gueldenstädt in his account of the jackal in Nov. Comm. Acad. Sc. Imp. Petrop. 1775, tom. xx. p. 449. See also another excellent account of the manners of this animal and of its play, in 'Land and Water,' October, 1869. Lieut. Annesley, R. A., has also communicated to me some particulars with respect to the jackal. I have made many inquiries about wolves and jackals in the Zoological Gardens, and have observed them for myself.

[5] 'Land and Water,' November 6, 1869.

quired, through the principle of antithesis, certain expressive gestures; for it is not probable that these animals, confined in cages, should have learnt them by imitating dogs.

Cats.—I have already described the actions of a cat

FIG. 15.—Cat terrified at a dog. From life, by Mr. Wood.

(fig. 9), when feeling savage and not terrified. She assumes a crouching attitude and occasionally protrudes her fore-feet, with the claws exserted ready for striking.

The tail is extended, being curled or lashed from side to side. The hair is not erected—at least it was not so in the few cases observed by me. The ears are drawn closely backwards and the teeth are shown. Low savage growls are uttered. We can understand why the attitude assumed by a cat when preparing to fight with another cat, or in any way greatly irritated, is so widely different from that of a dog approaching another dog with hostile intentions; for the cat uses her fore-feet for striking, and this renders a crouching position convenient or necessary. She is also much more accustomed than a dog to lie concealed and suddenly spring on her prey. No cause can be assigned with certainty for the tail being lashed or curled from side to side. This habit is common to many other animals—for instance, to the puma, when prepared to spring;[6] but it is not common to dogs, or to foxes, as I infer from Mr. St. John's account of a fox lying in wait and seizing a hare. We have already seen that some kinds of lizards and various snakes, when excited, rapidly vibrate the tips of their tails. It would appear as if, under strong excitement, there existed an uncontrollable desire for movement of some kind, owing to nerve-force being freely liberated from the excited sensorium; and that as the tail is left free, and as its movement does not disturb the general position of the body, it is curled or lashed about.

All the movements of a cat, when feeling affectionate, are in complete antithesis to those just described. She now stands upright, with slightly arched back, tail perpendicularly raised, and ears erected; and she rubs her cheeks and flanks against her master or mistress. The desire to rub something is so strong in cats under this state of mind, that they may often be seen rubbing

---

[6] Azara, 'Quadrupèdes du Paraquay,' 1801, tom. i. p. 136.

themselves against the legs of chairs or tables, or against door-posts. This manner of expressing affection probably originated through association, as in the case of dogs, from the mother nursing and fondling her young; and perhaps from the young themselves loving each other and playing together. Another and very different gesture, expressive of pleasure, has already been described, namely, the curious manner in which young and even old cats, when pleased, alternately protrude their fore-feet, with separated toes, as if pushing against and sucking their mother's teats. This habit is so far analogous to that of rubbing against something, that both apparently are derived from actions performed during the nursing period. Why cats should show affection by rubbing so much more than do dogs, though the latter delight in contact with their masters, and why cats only occasionally lick the hands of their friends, whilst dogs always do so, I cannot say. Cats cleanse themselves by licking their own coats more regularly than do dogs. On the other hand, their tongues seem less well fitted for the work than the longer and more flexible tongues of dogs.

Cats, when terrified, stand at full height, and arch their backs in a well-known and ridiculous fashion. They spit, hiss, or growl. The hair over the whole body, and especially on the tail, becomes erect. In the instances observed by me the basal part of the tail was held upright, the terminal part being thrown on one side; but sometimes the tail (see fig. 15) is only a little raised, and is bent almost from the base to one side. The ears are drawn back, and the teeth exposed. When two kittens are playing together, the one often thus tries to frighten the other. From what we have seen in former chapters, all the above points of expression are intelligible, except the extreme arching of the back. I am inclined to believe that, in the same manner as many birds, whilst they ruffle their feathers, spread out their wings and tail, to make themselves look as big as possible, so cats stand upright at their full height, arch their backs, often raise the basal part of the tail, and erect their hair, for the same purpose. The lynx, when attacked, is said to arch its back, and is thus figured by Brehm. But the keepers in the Zoological Gardens have never seen any tendency to this action in the larger feline animals, such as tigers, lions, &c.; and these have little cause to be afraid of any other animal.

Cats use their voices much as a means of expression, and they utter, under various emotions and desires, at least six or seven different sounds. The purr of satisfaction, which is made during both inspiration and expiration, is one of the most curious. The puma, cheetah, and ocelot likewise purr; but the tiger, when pleased, "emits a peculiar short snuffle, accompanied by the closure of the eyelids."[7] It is said that the lion, jaguar, and leopard, do not purr.

Horses.—Horses when savage draw their ears closely back, protrude their heads, and partially uncover their incisor teeth, ready for biting. When inclined to kick behind, they generally, through habit, draw back their ears; and their eyes are turned backwards in a peculiar manner.[8] When pleased, as when some coveted food is brought to them in the stable, they raise and draw in their heads, prick their ears, and looking intently towards their friend, often whinny. Impatience is expressed by pawing the ground.

[7] 'Land and Water,' 1867, p. 657. See also Azara on the Puma, in the work above quoted.

[8] Sir C. Bell, 'Anatomy of Expression,' 3rd edit. p. 123. See also p. 126, on horses not breathing through their mouths, with reference to their distended nostrils.

The actions of a horse when much startled are highly expressive. One day my horse was much frightened at a drilling machine, covered by a tarpaulin, and lying on an open field. He raised his head so high, that his neck became almost perpendicular; and this he did from habit, for the machine lay on a slope below, and could not have been seen with more distinctness through the raising of the head; nor if any sound had proceeded from it, could the sound have been more distinctly heard. His eyes and ears were directed intently forwards; and I could feel through the saddle the palpitations of his heart. With red dilated nostrils he snorted violently, and whirling round, would have dashed off at full speed, had I not prevented him. The distension of the nostrils is not for the sake of scenting the source of danger, for when a horse smells carefully at any object and is not alarmed, he does not dilate his nostrils. Owing to the presence of a valve in the throat, a horse when panting does not breathe through his open mouth, but through his nostrils; and these consequently have become endowed with great powers of expansion. This expansion of the nostrils, as well as the snorting, and the palpitations of the heart, are actions which have become firmly associated during a long series of generations with the emotion of terror; for terror has habitually led the horse to the most violent exertion in dashing away at full speed from the cause of danger.

Ruminants.—Cattle and sheep are remarkable from displaying in so slight a degree their emotions or sensations, excepting that of extreme pain. A bull when enraged exhibits his rage only by the manner in which he holds his lowered head, with distended nostrils, and by bellowing. He also often paws the ground; but this pawing seems quite different from that of an impatient horse, for when the soil is loose, he throws up clouds of dust. I believe that bulls act in this manner when irritated by flies, for the sake of driving them away. The wilder breeds of sheep and the chamois when startled stamp on the ground, and whistle through their noses; and this serves as a danger-signal to their comrades. The musk-ox of the Arctic regions, when encountered, likewise stamps on the ground.[9] How this stamping action arose I cannot conjecture; for from inquiries which I have made it does not appear that any of these animals fight with their fore-legs.

Some species of deer, when savage, display far more expression than do cattle, sheep, or goats, for, as has already been stated, they draw back their ears, grind their teeth, erect their hair, squeal, stamp on the ground, and brandish their horns. One day in the Zoological Gardens, the Formosan deer (*Cervus pseudaxis*) approached me in a curious attitude, with his muzzle raised high up, so that the horns were pressed back on his neck; the head being held rather obliquely. From the expression of his eye I felt sure that he was savage; he approached slowly, and as soon as he came close to the iron bars, he did not lower his head to butt at me, but suddenly bent it inwards, and struck his horns with great force against the railings. Mr. Bartlett informs me that some other species of deer place themselves in the same attitude when enraged.

Monkeys.—The various species and genera of monkeys express their feelings in many different ways; and this fact is interesting, as in some degree bearing on the question, whether the so-called races of man should be ranked as distinct species or varieties; for, as we shall

[9] 'Land and Water,' 1869, p. 152.

see in the following chapters, the different races of man express their emotions and sensations with remarkable uniformity throughout the world. Some of the expressive actions of monkeys are interesting in another way, namely from being closely analogous to those of man. As I have had no opportunity of observing any one species of the group under all circumstances, my miscellaneous remarks will be best arranged under different states of the mind.

Pleasure, joy, affection.—It is not possible to distinguish in monkeys, at least without more experience than I have had, the expression of pleasure or joy from that of affection. Young chimpanzees make a kind of barking noise, when pleased by the return of any one to whom they are attached. When this noise, which the keepers call a laugh, is uttered, the lips are protruded; but so they are under various other emotions. Nevertheless I could perceive that when they were pleased the form of the lips differed a little from that assumed when they were angered. If a young chimpanzee be tickled—and the armpits are particularly sensitive to tickling, as in the case of our children,—a more decided chuckling or laughing sound is uttered; though the laughter is sometimes noiseless. The corners of the mouth are then drawn backwards; and this sometimes causes the lower eyelids to be slightly wrinkled. But this wrinkling, which is so characteristic of our own laughter, is more plainly seen in some other monkeys. The teeth in the upper jaw in the chimpanzee are not exposed when they utter their laughing noise, in which respect they differ from us. But their eyes sparkle and grow brighter, as Mr. W. L. Martin,[10] who has particularly attended to their expression, states.

[10] 'Natural History of Mammalia,' 1841, vol. i. pp. 383, 410.

Young Orangs, when tickled, likewise grin and make a chuckling sound; and Mr. Martin says that their eyes grow brighter. As soon as their laughter ceases, an expression may be detected passing over their faces, which, as Mr. Wallace remarked to me, may be called a smile. I have also noticed something of the same kind with the chimpanzee. Dr. Duchenne—and I cannot quote a better authority—informs me that he kept a very tame monkey in his house for a year; and when he gave it during meal-times some choice delicacy, he observed that the corners of its mouth were slightly raised; thus an expression of satisfaction, partaking of the nature of an incipient smile, and resembling that often seen on the face of man, could be plainly perceived in this animal.

The *Cebus azaræ*,[11] when rejoiced at again seeing a beloved person, utters a peculiar tittering (*kichernden*) sound. It also expresses agreeable sensations, by drawing back the corners of its mouth, without producing any sound. Rengger calls this movement laughter, but it would be more appropriately called a smile. The form of the mouth is different when either pain or terror is expressed, and high shrieks are uttered. Another species of *Cebus* in the Zoological Gardens (*C. hypoleucus*) when pleased, makes a reiterated shrill note, and likewise draws back the corners of its mouth, apparently through the contraction of the same muscles as with us. So does the Barbary ape (*Inuus ecaudatus*) to an extraordinary degree; and I observed in this monkey that the skin of the lower eyelids then became much wrinkled. At the same time it rapidly moved its lower jaw or lips in a spasmodic manner, the teeth being exposed; but the noise produced was hardly more distinct than that which

[11] Rengger ('Säugetheire von Paraquay', 1830, s. 46) kept these monkeys in confinement for seven years in their native country of Paraguay.

we sometimes call silent laughter. Two of the keepers affirmed that this slight sound was the animal's laughter, and when I expressed some doubt on this head (being at the time quite inexperienced), they made it attack or rather threaten a hated Entellus monkey, living in the same compartment. Instantly the whole expression of the face of the Inuus changed; the mouth was opened much more widely, the canine teeth were more fully exposed, and a hoarse barking noise was uttered.

The Anubis baboon (*Cynocephalus anubis*) was first insulted and put into a furious rage, as was easily done, by his keeper, who then made friends with him and shook hands. As the reconciliation was effected the baboon rapidly moved up and down his jaws and lips, and looked pleased. When we laugh heartily, a similar movement, or quiver, may be observed more or less distinctly in our jaws; but with man the muscles of the chest are more particularly acted on, whilst with this baboon, and with some other monkeys, it is the muscles of the jaws and lips which are spasmodically affected.

I have already had occasion to remark on the curious manner in which two or three species of Macacus and the *Cynopithecus niger* draw back their ears and utter a slight jabbering noise, when they are pleased by being caressed. With the Cynopithecus (fig. 17), the corners of the mouth are at the same time drawn backwards and upwards, so that the teeth are exposed. Hence this expression would never be recognized by a stranger as one of pleasure. The crest of long hairs on the forehead is depressed, and apparently the whole skin of the head drawn backwards. The eyebrows are thus raised a little, and the eyes assume a staring appearance. The lower eyelids also become slightly wrinkled; but this wrinkling is not conspicuous, owing to the permanent transverse furrows on the face.

Painful emotions and sensations.—With monkeys the expression of slight pain, or of any painful emotion, such as grief, vexation, jealousy, &c., is not easily distinguished from that of moderate anger; and these states of mind readily and quickly pass into each other. Grief, however, with some species is certainly exhibited by weeping. A woman, who sold a monkey to the Zoological Society, believed to have come from Borneo (*Macacus maurus* or *M. inornatus* of Gray), said that it often cried; and Mr. Bartlett, as well as the keeper Mr. Sutton, have repeatedly seen it, when grieved, or even when much pitied, weeping so copiously that the tears rolled down its cheeks. There is, however, something strange about this case, for two specimens subsequently kept in the Gardens, and believed to be the same species, have never been seen to weep, though they were carefully observed by the keeper and myself when much distressed and loudly screaming. Rengger states [12] that the eyes of the *Cebus azaræ* fill with tears, but not sufficiently to overflow, when it is prevented getting some much desired object, or is much frightened. Humboldt also asserts that the eyes of the *Callithrix sciureus* "instantly fill with tears when it is seized with fear;" but when this pretty little monkey in the Zoological Gardens was teased, so as to cry out loudly, this did not occur. I do not, however, wish to throw the least doubt on the accuracy of Humboldt's statement.

The appearance of dejection in young orangs and chimpanzees, when out of health, is as plain and almost as pathetic as in the case of our children. This state of mind and body is shown by their listless movements, fallen countenances, dull eyes, and changed complexion.

[12] Rengger, ibid. s. 46. Humboldt, 'Personal Narrative,' Eng. translat. vol. iv. p. 527.

FIG. 16.—*Cynopithecus niger* in a placid condition.
Drawn from life by Mr. Wolf.

FIG. 17.—The same, when pleased by being caressed.

Anger.—This emotion is often exhibited by many kinds of monkeys, and is expressed, as Mr. Martin remarks,[13] in many different ways. "Some species, when irritated, pout the lips, gaze with a fixed and savage glare on their foe, and make repeated short starts as if about to spring forward, uttering at the same time inward guttural sounds. Many display their anger by suddenly advancing, making abrupt starts, at the same time opening the mouth and pursing up the lips, so as to conceal the teeth, while the eyes are daringly fixed on the enemy, as if in savage defiance. Some again, and principally the long-tailed monkeys, or Guenons, display their teeth, and accompany their malicious grins with a sharp, abrupt, reiterated cry." Mr. Sutton confirms the statement that some species uncover their teeth when enraged, whilst others conceal them by the protrusion of their lips; and some kinds draw back their ears. The *Cynopithecus niger*, lately referred to, acts in this manner, at the same time depressing the crest of hair on its forehead, and showing its teeth; so that the movements of the features from anger are nearly the same as those from pleasure; and the two expressions can be distinguished only by those familiar with the animal.

Baboons often show their passion and threaten their enemies in a very odd manner, namely, by opening their mouths widely as in the act of yawning. Mr. Bartlett has often seen two baboons, when first placed in the same compartment, sitting opposite to each other and thus alternately opening their mouths; and this action seems frequently to end in a real yawn. Mr. Bartlett believes that both animals wish to show to each other that they are provided with a formidable set of teeth, as is undoubtedly the case. As I could hardly credit the

---

[13] Nat. Hist. of Mammalia, 1841, p. 351.

reality of this yawning gesture, Mr. Bartlett insulted an old baboon and put him into a violent passion; and he almost immediately thus acted. Some species of Macacus and of Cercopithecus[14] behave in the same manner. Baboons likewise show their anger, as was observed by Brehm with those which he kept alive in Abyssinia, in another manner, namely, by striking the ground with one hand, "like an angry man striking the table with his fist." I have seen this movement with the baboons in the Zoological Gardens; but sometimes the action seems rather to represent the searching for a stone or other object in their beds of straw.

Mr. Sutton has often observed the face of the *Macacus rhesus*, when much enraged, growing red. As he was mentioning this to me, another monkey attacked a *rhesus*, and I saw its face redden as plainly as that of a man in a violent passion. In the course of a few minutes, after the battle, the face of this monkey recovered its natural tint. At the same time that the face reddened, the naked posterior part of the body, which is always red, seemed to grow still redder; but I cannot positively assert that this was the case. When the Mandrill is in any way excited, the brilliantly coloured, naked parts of the skin are said to become still more vividly coloured.

With several species of baboons the ridge of the forehead projects much over the eyes, and is studded with a few long hairs, representing our eyebrows. These animals are always looking about them, and in order to look upwards they raise their eyebrows. They have thus, as it would appear, acquired the habit of frequently moving their eyebrows. However this may be, many kinds of monkeys, especially the baboons, when angered

---

[14] Brehm, 'Thierleben,' B. i. s. 84. On baboons striking the ground, s. 61.

or in any way excited, rapidly and incessantly move their eyebrows up and down, as well as the hairy skin of their foreheads.[15] As we associate in the case of man the raising and lowering of the eyebrows with definite states of the mind, the almost incessant movement of the eyebrows by monkeys gives them a senseless expression. I once observed a man who had a trick of continually raising his eyebrows without any corresponding emotion, and this gave to him a foolish appearance; so it is with some persons who keep the corners of their mouths a little drawn backwards and upwards, as if by an incipient smile, though at the time they are not amused or pleased.

A young orang, made jealous by her keeper attending to another monkey, slightly uncovered her teeth, and, uttering a peevish noise like *tish-shist*, turned her back on him. Both orangs and chimpanzees, when a little more angered, protrude their lips greatly, and make a harsh barking noise. A young female chimpanzee, in a violent passion, presented a curious resemblance to a child in the same state. She screamed loudly with widely open mouth, the lips being retracted so that the teeth were fully exposed. She threw her arms wildly about, sometimes clasping them over her head. She rolled on the ground, sometimes on her back, sometimes on her belly, and bit everything within reach. A young gibbon (*Hylobates syndactylus*) in a passion has been described[16] as behaving in almost exactly the same manner.

The lips of young orangs and chimpanzees are protruded, sometimes to a wonderful degree, under various circumstances. They act thus, not only when slightly angered, sulky, or disappointed, but when alarmed at

---

[15] Brehm remarks ('Thierleben,' s. 68) that the eyebrows of the *Inuus ecaudatus* are frequently moved up and down when the animal is angered.

[16] G. Bennett, 'Wanderings in New South Wales,' &c. vol. ii. 1834, p. 153.

FIG. 18.—Chimpanzee disappointed and sulky. Drawn from life by Mr. Wood.

anything—in one instance, at the sight of a turtle,[17]—and likewise when pleased. But neither the degree of protrusion nor the shape of the mouth is exactly the same, as I believe, in all cases; and the sounds which are then uttered are different. The accompanying drawing represents a chimpanzee made sulky by an orange having been offered him, and then taken away. A similar protrusion or pouting of the lips, though to a much slighter degree, may be seen in sulky children.

Many years ago, in the Zoological Gardens, I placed a looking-glass on the floor before two young orangs, who, as far as it was known, had never before seen one. At first they gazed at their own images with the most steady surprise, and often changed their point of view. They then approached close and protruded their lips towards the image, as if to kiss it, in exactly the same manner as they had previously done towards each other, when first placed, a few days before, in the same room. They next made all sorts of grimaces, and put themselves in various attitudes before the mirror; they pressed and rubbed the surface; they placed their hands at different distances behind it; looked behind it; and finally seemed almost frightened, started a little, became cross, and refused to look any longer.

When we try to perform some little action which is difficult and requires precision, for instance, to thread a needle, we generally close our lips firmly, for the sake, I presume, of not disturbing our movements by breathing; and I noticed the same action in a young Orang. The poor little creature was sick, and was amusing itself by trying to kill the flies on the window-panes with its

---

[17] W. L. Martin, Nat. Hist. of Mamm. Animals, 1841, p. 405.

knuckles; this was difficult as the flies buzzed about, and at each attempt the lips were firmly compressed, and at the same time slightly protruded.

Although the countenances, and more especially the gestures, of orangs and chimpanzees are in some respects highly expressive, I doubt whether on the whole they are so expressive as those of some other kinds of monkeys. This may be attributed in part to their ears being immovable, and in part to the nakedness of their eyebrows, of which the movements are thus rendered less conspicuous. When, however, they raise their eyebrows their foreheads become, as with us, transversely wrinkled. In comparison with man, their faces are inexpressive, chiefly owing to their not frowning under any emotion of the mind—that is, as far as I have been able to observe, and I carefully attended to this point. Frowning, which is one of the most important of all the expressions in man, is due to the contraction of the corrugators by which the eyebrows are lowered and brought together, so that vertical furrows are formed on the forehead. Both the orang and chimpanzee are said [18] to possess this muscle, but it seems rarely brought into action, at least in a conspicuous manner. I made my hands into a sort of cage, and placing some tempting fruit within, allowed both a young orang and chimpanzee to try their utmost to get it out; but although they grew rather cross, they showed not a trace of a frown. Nor was there any frown when they were enraged. Twice I took two chimpanzees from their rather dark room suddenly into bright sunshine, which would certainly have caused us to frown; they blinked and winked their eyes, but only

[18] Prof. Owen on the Orang, Proc. Zool. Soc. 1830, p. 28. On the Chimpanzee, see Prof. Macalister, in Annals and Mag. of Nat. Hist. vol. vii. 1871, p. 342, who states that the *corrugator supercilii* is inseparable from the *orbicularis palpebrarum*.

once did I see a very slight frown. On another occasion, I tickled the nose of a chimpanzee with a straw, and as it crumpled up its face, slight vertical furrows appeared between the eyebrows. I have never seen a frown on the forehead of the orang.

The gorilla, when enraged, is described as erecting its crest of hair, throwing down its under lip, dilating its nostrils, and uttering terrific yells. Messrs. Savage and Wyman [19] state that the scalp can be freely moved backwards and forwards, and that when the animal is excited it is strongly contracted; but I presume that they mean by this latter expression that the scalp is lowered; for they likewise speak of the young chimpanzee, when crying out, "as having the eyebrows strongly contracted." The great power of movement in the scalp of the gorilla, of many baboons and other monkeys, deserves notice in relation to the power possessed by some few men, either through reversion or persistence, of voluntarily moving their scalps.[20]

Astonishment, Terror.—A living fresh-water turtle was placed at my request in the same compartment in the Zoological Gardens with many monkeys; and they showed unbounded astonishment, as well as some fear. This was displayed by their remaining motionless, staring intently with widely opened eyes, their eyebrows being often moved up and down. Their faces seemed somewhat lengthened. They occasionally raised themselves on their hind-legs to get a better view. They often retreated a few feet, and then turning their heads over one shoulder, again stared intently. It was curious to observe how much less afraid they were of the turtle than of a living snake which I had formerly placed in

---

[19] Boston Journal of Nat. Hist. 1845--47, vol. v. p. 423. On the Chimpanzee, ibid. 1843--44, vol. iv. p. 365.

[20] See on this subject, 'Descent of Man,' vol. i. p. 20.

their compartment;[21] for in the course of a few minutes some of the monkeys ventured to approach and touch the turtle. On the other hand, some of the larger baboons were greatly terrified, and grinned as if on the point of screaming out. When I showed a little dressed-up doll to the *Cynopithecus niger*, it stood motionless, stared intently with widely opened eyes, and advanced its ears a little forwards. But when the turtle was placed in its compartment, this monkey also moved its lips in an odd, rapid, jabbering manner, which the keeper declared was meant to conciliate or please the turtle.

I was never able clearly to perceive that the eyebrows of astonished monkeys were kept permanently raised, though they were frequently moved up and down. Attention, which precedes astonishment, is expressed by man by a slight raising of the eyebrows; and Dr. Duchenne informs me that when he gave to the monkey formerly mentioned some quite new article of food, it elevated its eyebrows a little, thus assuming an appearance of close attention. It then took the food in its fingers, and, with lowered or rectilinear eyebrows, scratched, smelt, and examined it,—an expression of reflection being thus exhibited. Sometimes it would throw back its head a little, and again with suddenly raised eyebrows re-examine and finally taste the food.

In no case did any monkey keep its mouth open when it was astonished. Mr. Sutton observed for me a young orang and chimpanzee during a considerable length of time; and however much they were astonished, or whilst listening intently to some strange sound, they did not keep their mouths open. This fact is surprising, as with

---

[21] 'Descent of Man,' vol. i. p. 43.

mankind hardly any expression is more general than a widely open mouth under the sense of astonishment. As far as I have been able to observe, monkeys breathe more freely through their nostrils than men do; and this may account for their not opening their mouths when they are astonished; for, as we shall see in a future chapter, man apparently acts in this manner when startled, at first for the sake of quickly drawing a full inspiration, and afterwards for the sake of breathing as quietly as possible.

Terror is expressed by many kinds of monkeys by the utterance of shrill screams; the lips being drawn back, so that the teeth are exposed. The hair becomes erect, especially when some anger is likewise felt. Mr. Sutton has distinctly seen the face of the *Macacus rhesus* grow pale from fear. Monkeys also tremble from fear; and sometimes they void their excretions. I have seen one which, when caught, almost fainted from an excess of terror.

Sufficient facts have now been given with respect to the expressions of various animals. It is impossible to agree with Sir C. Bell when he says [22] that "the faces of animals seem chiefly capable of expressing rage and fear;" and again, when he says that all their expressions "may be referred, more or less plainly, to their acts of volition or necessary instincts." He who will look at a dog preparing to attack another dog or a man, and at the same animal when caressing his master, or will watch the countenance of a monkey when insulted, and when fondled by his keeper, will be forced to admit that the movements of their features and their gestures are almost as expressive as those of man. Although no explanation

[22] 'Anatomy of Expression,' 3rd edit. 1844, pp. 138, 121.

can be given of some of the expressions in the lower animals, the greater number are explicable in accordance with the three principles given at the commencement of the first chapter.

CHAPTER VI.

SPECIAL EXPRESSIONS OF MAN: SUFFERING AND WEEPING.

The screaming and weeping of infants—Forms of features—Age at which weeping commences—The effects of habitual restraint on weeping—Sobbing—Cause of the contraction of the muscles round the eyes during screaming—Cause of the secretion of tears.

IN this and the following chapters the expressions exhibited by Man under various states of the mind will be described and explained, as far as lies in my power. My observations will be arranged according to the order which I have found the most convenient; and this will generally lead to opposite emotions and sensations succeeding each other.

Suffering of the body and mind: weeping.—I have already described in sufficient detail, in the third chapter, the signs of extreme pain, as shown by screams or groans, with the writhing of the whole body and the teeth clenched or ground together. These signs are often accompanied or followed by profuse sweating, pallor, trembling, utter prostration, or faintness. No suffering is greater than that from extreme fear or horror, but here a distinct emotion comes into play, and will be elsewhere considered. Prolonged suffering, especially of the mind, passes into low spirits, grief, dejection, and despair, and these states will be the subject of the follow-

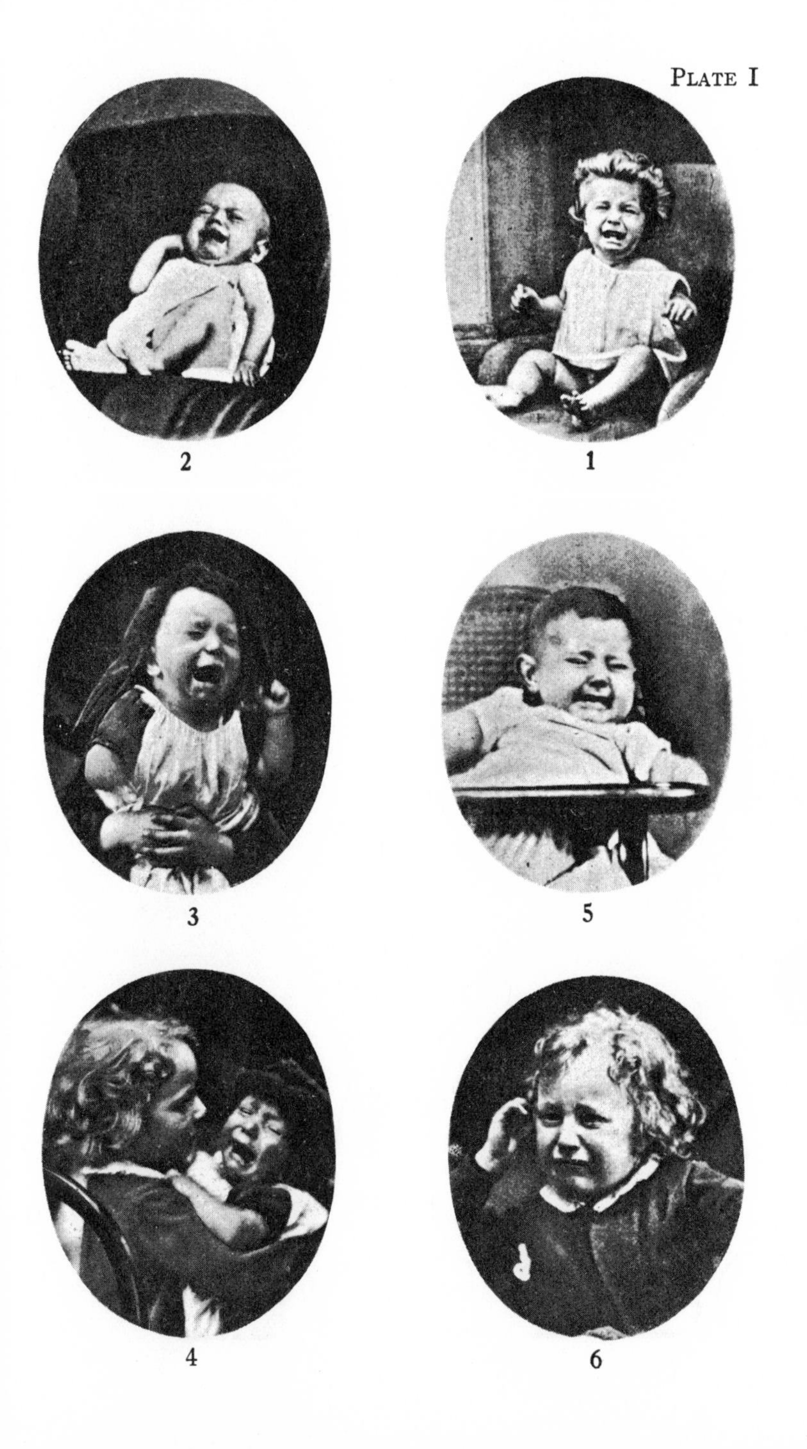
2
1
3
5
4
6

ing chapter. Here I shall almost confine myself to weeping or crying, more especially in children.

Infants, when suffering even slight pain, moderate hunger, or discomfort, utter violent and prolonged screams. Whilst thus screaming their eyes are firmly closed, so that the skin round them is wrinkled, and the forehead contracted into a frown. The mouth is widely opened with the lips retracted in a peculiar manner, which causes it to assume a squarish form; the gums or teeth being more or less exposed. The breath is inhaled almost spasmodically. It is easy to observe infants whilst screaming; but I have found photographs made by the instantaneous process the best means for observation, as allowing more deliberation. I have collected twelve, most of them made purposely for me; and they all exhibit the same general characteristics. I have, therefore, had six of them [1] (Plate I.) reproduced by the heliotype process.

The firm closing of the eyelids and consequent compression of the eyeball,—and this is a most important element in various expressions,—serves to protect the eyes from becoming too much gorged with blood, as will presently be explained in detail. With respect to the order in which the several muscles contract in firmly compressing the eyes, I am indebted to Dr. Langstaff, of Southampton, for some observations, which I have since repeated. The best plan for observing the order is to make a person first raise his eyebrows, and this produces transverse wrinkles across the forehead; and then very gradually to contract all the muscles round the eyes

---

[1] The best photographs in my collection are by Mr. Rejlander, of Victoria Street, London, and by Herr Kindermann, of Hamburg. Figs. 1, 3, 4, and 6 are by the former; and figs. 2 and 5, by the latter gentleman. Fig. 6 is given to show moderate crying in an older child.

with as much force as possible. The reader who is unacquainted with the anatomy of the face, ought to refer to p. 24, and look at the woodcuts 1 to 3. The corrugators of the brow (*corrugator supercilii*) seem to be the first muscles to contract; and these draw the eyebrows downwards and inwards towards the base of the nose, causing vertical furrows, that is a frown, to appear between the eyebrows; at the same time they cause the disappearance of the transverse wrinkles across the forehead. The orbicular muscles contract almost simultaneously with the corrugators, and produce wrinkles all round the eyes; they appear, however, to be enabled to contract with greater force, as soon as the contraction of the corrugators has given them some support. Lastly, the pyramidal muscles of the nose contract; and these draw the eyebrows and the skin of the forehead still lower down, producing short transverse wrinkles across the base of the nose.[2] For the sake of brevity these muscles will generally be spoken of as the orbiculars, or as those surrounding the eyes.

When these muscles are strongly contracted, those running to the upper lip[3] likewise contract and raise the upper lip. This might have been expected from the manner in which at least one of them, the *malaris*,

---

[2] Henle ('Handbuch d. Syst. Anat. 1858, B. i. s. 139) agrees with Duchenne that this is the effect of the contraction of the *pyramidalis nasi.*

[3] These consist of the *levator labii superioris alæque nasi*, the *levator labii proprius*, the *malaris*, and the *zygomaticus minor*, or little zygomatic. This latter muscle runs parallel to and above the great zygomatic, and is attached to the outer part of the upper lip. It is represented in fig. 2 (I. p. 24), but not in figs. 1 and 3. Dr. Duchenne first showed ('Mécanisme de la Physionomie Humaine,' Album, 1862, p. 39) the importance of the contraction of this muscle in the shape assumed by the features in crying. Henle considers the above-named muscles (excepting the *malaris*) as subdivisions of the *quadratus labii superioris.*

is connected with the orbiculars. Any one who will gradually contract the muscles round his eyes, will feel, as he increases the force, that his upper lip and the wings of his nose (which are partly acted on by one of the same muscles) are almost always a little drawn up. If he keeps his mouth firmly shut whilst contracting the muscles round the eyes, and then suddenly relaxes his lips, he will feel that the pressure on his eyes immediately increases. So again when a person on a bright, glaring day wishes to look at a distant object, but is compelled partially to close his eyelids, the upper lip may almost always be observed to be somewhat raised. The mouths of some very short-sighted persons, who are forced habitually to reduce the aperture of their eyes, wear from this same reason a grinning expression.

The raising of the upper lip draws upwards the flesh of the upper parts of the cheeks, and produces a strongly marked fold on each cheek,—the naso-labial fold,—which runs from near the wings of the nostrils to the corners of the mouth and below them. This fold or furrow may be seen in all the photographs, and is very characteristic of the expression of a crying child; though a nearly similar fold is produced in the act of laughing or smiling.[4]

---

[4] Although Dr. Duchenne has so carefully studied the contraction of the different muscles during the act of crying, and the furrows on the face thus produced, there seems to be something incomplete in his account; but what this is I cannot say. He has given a figure (Album, fig. 48) in which one half of the face is made, by galvanizing the proper muscles, to smile; whilst the other half is similarly made to begin crying. Almost all those (viz. nineteen out of twenty-one persons) to whom I showed the smiling half of the face instantly recognized the expression; but, with respect to the other half, only six persons out of twenty-one recognized it,—that is, if we accept such terms as "grief," "misery," "annoyance," as correct;—whereas, fifteen persons were ludicrously mistaken; some of them saying the face ex-

As the upper lip is much drawn up during the act of screaming, in the manner just explained, the depressor muscles of the angles of the mouth (see K in woodcuts 1 and 2) are strongly contracted in order to keep the mouth widely open, so that a full volume of sound may be poured forth. The action of these opposed muscles, above and below, tends to give to the mouth an oblong, almost squarish outline, as may be seen in the accompanying photographs. An excellent observer,[5] in describing a baby crying whilst being fed, says, "it made its mouth like a square, and let the porridge run out at all four corners." I believe, but we shall return to this point in a future chapter, that the depressor muscles of the angles of the mouth are less under the separate control of the will than the adjoining muscles; so that if a young child is only doubtfully inclined to cry, this muscle is generally the first to contract, and is the last to cease contracting. When older children commence crying, the muscles which run to the upper lip are often the first to contract; and this may perhaps be due to older children not having so strong a tendency to scream loudly, and consequently to keep their mouths widely

pressed "fun," "satisfaction," "cunning," "disgust," &c. We may infer from this that there is something wrong in the expression. Some of the fifteen persons may, however, have been partly misled by not expecting to see an old man crying, and by tears not being secreted. With respect to another figure by Dr. Duchenne (fig. 49), in which the muscles of half the face are galvanized in order to represent a man beginning to cry, with the eyebrow on the same side rendered oblique, which is characteristic of misery, the expression was recognized by a greater proportional number of persons. Out of twenty-three persons, fourteen answered correctly, "sorrow," "distress," "grief," "just going to cry," "endurance of pain," &c. On the other hand, nine persons either could form no opinion or were entirely wrong, answering, "cunning leer," "jocund," "looking at an intense light," "looking at a distant object," &c.

[5] Mrs. Gaskell, 'Mary Barton,' new edit. p. 84.

open; so that the above-named depressor muscles are not brought into such strong action.

With one of my own infants, from his eighth day and for some time afterwards, I often observed that the first sign of a screaming-fit, when it could be observed coming on gradually, was a little frown, owing to the contraction of the corrugators of the brows; the capillaries of the naked head and face becoming at the same time reddened with blood. As soon as the screaming-fit actually began, all the muscles round the eyes were strongly contracted, and the mouth widely opened in the manner above described; so that at this early period the features assumed the same form as at a more advanced age.

Dr. Piderit [6] lays great stress on the contraction of certain muscles which draw down the nose and narrow the nostrils, as eminently characteristic of a crying expression. The *depressores anguli oris*, as we have just seen, are usually contracted at the same time, and they indirectly tend, according to Dr. Duchenne, to act in this same manner on the nose. With children having bad colds a similar pinched appearance of the nose may be noticed, which is at least partly due, as remarked to me by Dr. Langstaff, to their constant snuffling, and the consequent pressure of the atmosphere on the two sides. The purpose of this contraction of the nostrils by children having bad colds, or whilst crying, seems to be to check the downward flow of the mucus and tears, and to prevent these fluids spreading over the upper lip.

After a prolonged and severe screaming-fit, the scalp, face, and eyes are reddened, owing to the return of the blood from the head having been impeded by the violent expiratory efforts; but the redness of the stimulated

[6] 'Mimik und Physiognomik,' 1867, s. 102. Duchenne, Mécanisme de la Phys. Humaine, Album, p. 34.

eyes is chiefly due to the copious effusion of tears. The various muscles of the face which have been strongly contracted, still twitch a little, and the upper lip is still slightly drawn up or everted,[7] with the corners of the mouth still a little drawn downwards. I have myself felt, and have observed in other grown-up persons, that when tears are restrained with difficulty, as in reading a pathetic story, it is almost impossible to prevent the various muscles, which with young children are brought into strong action during their screaming-fits, from slightly twitching or trembling.

Infants whilst young do not shed tears or weep, as is well known to nurses and medical men. This circumstance is not exclusively due to the lacrymal glands being as yet incapable of secreting tears. I first noticed this fact from having accidentally brushed with the cuff of my coat the open eye of one of my infants, when seventy-seven days old, causing this eye to water freely; and though the child screamed violently, the other eye remained dry, or was only slightly suffused with tears. A similar slight effusion occurred ten days previously in both eyes during a screaming-fit. The tears did not run over the eyelids and roll down the cheeks of this child, whilst screaming badly, when 122 days old. This first happened 17 days later, at the age of 139 days. A few other children have been observed for me, and the period of free weeping appears to be very variable. In one case, the eyes became slightly suffused at the age of only 20 days; in another, at 62 days. With two other children, the tears did *not* run down the face at the ages of 84 and 110 days; but in a third child they did run down at the age of 104 days. In one instance, as I was positively assured, tears ran down at the unusually early

---

[7] Dr. Duchenne makes this remark, ibid. p. 39.

age of 42 days. It would appear as if the lacrymal glands required some practice in the individual before they are easily excited into action, in somewhat the same manner as various inherited consensual movements and tastes require some exercise before they are fixed and perfected. This is all the more likely with a habit like weeping, which must have been acquired since the period when man branched off from the common progenitor of the genus Homo and of the non-weeping anthropomorphous apes.

The fact of tears not being shed at a very early age from pain or any mental emotion is remarkable, as, later in life, no expression is more general or more strongly marked than weeping. When the habit has once been acquired by an infant, it expresses in the clearest manner suffering of all kinds, both bodily pain and mental distress, even though accompanied by other emotions, such as fear or rage. The character of the crying, however, changes at a very early age, as I noticed in my own infants,—the passionate cry differing from that of grief. A lady informs me that her child, nine months old, when in a passion screams loudly, but does not weep; tears, however, are shed when she is punished by her chair being turned with its back to the table. This difference may perhaps be attributed to weeping being restrained, as we shall immediately see, at a more advanced age, under most circumstances excepting grief; and to the influence of such restraint being transmitted to an earlier period of life, than that at which it was first practised.

With adults, especially of the male sex, weeping soon ceases to be caused by, or to express, bodily pain. This may be accounted for by its being thought weak and unmanly by men, both of civilized and barbarous races, to exhibit bodily pain by any outward sign. With this exception, savages weep copiously from very slight causes, of which fact Sir J. Lubbock[8] has collected instances. A New Zealand chief "cried like a child because the sailors spoilt his favourite cloak by powdering it with flour." I saw in Tierra del Fuego a native who had lately lost a brother, and who alternately cried with hysterical violence, and laughed heartily at anything which amused him. With the civilized nations of Europe there is also much difference in the frequency of weeping. Englishmen rarely cry, except under the pressure of the acutest grief; whereas in some parts of the Continent the men shed tears much more readily and freely.

The insane notoriously give way to all their emotions with little or no restraint; and I am informed by Dr. J. Crichton Browne, that nothing is more characteristic of simple melancholia, even in the male sex, than a tendency to weep on the slightest occasions, or from no cause. They also weep disproportionately on the occurrence of any real cause of grief. The length of time during which some patients weep is astonishing, as well as the amount of tears which they shed. One melancholic girl wept for a whole day, and afterwards confessed to Dr. Browne, that it was because she remembered that she had once shaved off her eyebrows to promote their growth. Many patients in the asylum sit for a long time rocking themselves backwards and forwards; "and if spoken to, they stop their movements, purse up their eyes, depress the corners of the mouth, and burst out crying." In some of these cases, the being spoken to or kindly greeted appears to suggest some fanciful and sorrowful notion; but in other cases an effort of any kind excites weeping, independently of any sorrowful idea. Patients suffering from acute mania likewise have parox-

---

[8] 'The Origin of Civilization,' 1870, p. 355.

ysms of violent crying or blubbering, in the midst of their incoherent ravings. We must not, however, lay too much stress on the copious shedding of tears by the insane, as being due to the lack of all restraint; for certain brain-diseases, as hemiplegia, brain-wasting, and senile decay, have a special tendency to induce weeping. Weeping is common in the insane, even after a complete state of fatuity has been reached and the power of speech lost. Persons born idiotic likewise weep;[9] but it is said that this is not the case with cretins.

Weeping seems to be the primary and natural expression, as we see in children, of suffering of any kind, whether bodily pain short of extreme agony, or mental distress. But the foregoing facts and common experience show us that a frequently repeated effort to restrain weeping, in association with certain states of the mind, does much in checking the habit. On the other hand, it appears that the power of weeping can be increased through habit; thus the Rev. R. Taylor,[10] who long resided in New Zealand, asserts that the women can voluntarily shed tears in abundance; they meet for this purpose to mourn for the dead, and they take pride in crying "in the most affecting manner."

A single effort of repression brought to bear on the lacrymal glands does little, and indeed seems often to lead to an opposite result. An old and experienced physician told me that he had always found that the only means to check the occasional bitter weeping of ladies who consulted him, and who themselves wished to desist, was earnestly to beg them not to try, and to assure

---

[9] See, for instance, Mr. Marshall's account of an idiot in Philosoph. Transact. 1864, p. 526. With respect to cretins, see Dr. Piderit, 'Mimik und Physiognomik,' 1867, s. 61.

[10] 'New Zealand and its Inhabitants,' 1855, p. 175.

them that nothing would relieve them so much as prolonged and copious crying.

The screaming of infants consists of prolonged expirations, with short and rapid, almost spasmodic inspirations, followed at a somewhat more advanced age by sobbing. According to Gratiolet,[11] the glottis is chiefly affected during the act of sobbing. This sound is heard "at the moment when the inspiration conquers the resistance of the glottis, and the air rushes into the chest." But the whole act of respiration is likewise spasmodic and violent. The shoulders are at the same time generally raised, as by this movement respiration is rendered easier. With one of my infants, when seventy-seven days old, the inspirations were so rapid and strong that they approached in character to sobbing; when 138 days old I first noticed distinct sobbing, which subsequently followed every bad crying-fit. The respiratory movements are partly voluntary and partly involuntary, and I apprehend that sobbing is at least in part due to children having some power to command after early infancy their vocal organs and to stop their screams, but from having less power over their respiratory muscles, these continue for a time to act in an involuntary or spasmodic manner, after having been brought into violent action. Sobbing seems to be peculiar to the human species; for the keepers in the Zoological Gardens assure me that they have never heard a sob from any kind of monkey; though monkeys often scream loudly whilst being chased and caught, and then pant for a long time. We thus see that there is a close analogy between sobbing and the free shedding of tears; for with children, sobbing does not commence during early infancy, but afterwards comes on rather suddenly and

---

[11] 'De la Physionomie,' 1865, p. 126.

then follows every bad crying-fit, until the habit is checked with advancing years.

On the cause of the contraction of the muscles round the eyes during screaming.—We have seen that infants and young children, whilst screaming, invariably close their eyes firmly, by the contraction of the surrounding muscles, so that the skin becomes wrinkled all around. With older children, and even with adults, whenever there is violent and unrestrained crying, a tendency to the contraction of these same muscles may be observed; though this is often checked in order not to interfere with vision.

Sir C. Bell explains [12] this action in the following manner:—"During every violent act of expiration, whether in hearty laughter, weeping, coughing, or sneezing, the eyeball is firmly compressed by the fibres of the orbicularis; and this is a provision for supporting and defending the vascular system of the interior of the eye from a retrograde impulse communicated to the blood in the veins at that time. When we contract the chest and expel the air, there is a retardation of the blood in the veins of the neck and head; and in the more powerful acts of expulsion, the blood not only distends the vessels, but is even regurgitated into the minute branches. Were the eye not properly compressed at that time, and a resistance given to the shock, irreparable injury might be inflicted on the delicate textures of the interior of the eye." He further adds, "If we separate the eyelids of a child to examine the eye, while it cries and struggles with passion, by taking off the natural support to the vascular system of the eye, and means of

---

[12] 'The Anatomy of Expression,' 1844, p. 106. See also his paper in the 'Philosophical Transactions,' 1822, p. 284, ibid. 1823, pp. 166 and 289. Also 'The Nervous System of the Human Body,' 3rd edit. 1836, p. 175.

guarding it against the rush of blood then occurring, the conjunctiva becomes suddenly filled with blood, and the eyelids everted."

Not only are the muscles round the eyes strongly contracted, as Sir C. Bell states and as I have often observed, during screaming, loud laughter, coughing, and sneezing, but during several other analogous actions. A man contracts these muscles when he violently blows his nose. I asked one of my boys to shout as loudly as he possibly could, and as soon as he began, he firmly contracted his orbicular muscles; I observed this repeatedly, and on asking him why he had every time so firmly closed his eyes, I found that he was quite unaware of the fact: he had acted instinctively or unconsciously.

It is not necessary, in order to lead to the contraction of these muscles, that air should actually be expelled from the chest; it suffices that the muscles of the chest and abdomen should contract with great force, whilst by the closure of the glottis no air escapes. In violent vomiting or retching the diaphragm is made to descend by the chest being filled with air; it is then held in this position by the closure of the glottis, "as well as by the contraction of its own fibres." [13] The abdominal muscles now contract strongly upon the stomach, its proper muscles likewise contracting, and the contents are thus ejected. During each effort of vomiting "the head becomes greatly congested, so that the features are red and swollen, and the large veins of the face and temples visibly dilated." At the same time, as I know from observation, the muscles round the eyes are strongly contracted. This is likewise the case when the abdominal muscles

[13] See Dr. Brinton's account of the act of vomiting, in Todd's Cyclop. of Anatomy and Physiology, 1859, vol. v. Supplement, p. 318.

act downwards with *unusual* force in expelling the contents of the intestinal canal.

The greatest exertion of the muscles of the body, if those of the chest are not brought into strong action in expelling or compressing the air within the lungs, does not lead to the contraction of the muscles round the eyes. I have observed my sons using great force in gymnastic exercises, as in repeatedly raising their suspended bodies by their arms alone, and in lifting heavy weights from the ground, but there was hardly any trace of contraction in the muscles round the eyes.

As the contraction of these muscles for the protection of the eyes during violent expiration is indirectly, as we shall hereafter see, a fundamental element in several of our most important expressions, I was extremely anxious to ascertain how far Sir C. Bell's view could be substantiated. Professor Donders, of Utrecht,[14] well known as one of the highest authorities in Europe on vision and on the structure of the eye, has most kindly undertaken for me this investigation with the aid of the many ingenious mechanisms of modern science, and has published the results.[15] He shows that during violent expiration the external, the intra-ocular, and the retro-ocular vessels of the eye are all affected in two ways, namely by the increased pressure of the blood in the arteries, and by the return of the blood in the veins

---

[14] I am greatly indebted to Mr. Bowman for having introduced me to Prof. Donders, and for his aid in persuading this great physiologist to undertake the investigation of the present subject. I am likewise much indebted to Mr. Bowman for having given me, with the utmost kindness, information on many points.

[15] This memoir first appeared in the 'Nederlandsch Archief voor Genees en Natuurkunde,' Deel 5, 1870. It has been translated by Dr. W. D. Moore, under the title of "On the Action of the Eyelids in determination of Blood from expiratory effort," in 'Archives of Medicine,' edited by Dr. L. S. Beale, 1870, vol. v. p. 20.

being impeded. It is, therefore, certain that both the arteries and the veins of the eye are more or less distended during violent expiration. The evidence in detail may be found in Professor Donders' valuable memoir. We see the effects on the veins of the head, in their prominence, and in the purple colour of the face of a man who coughs violently from being half choked. I may mention, on the same authority, that the whole eye certainly advances a little during each violent expiration. This is due to the dilatation of the retro-ocular vessels, and might have been expected from the intimate connection of the eye and brain; the brain being known to rise and fall with each respiration, when a portion of the skull has been removed; and as may be seen along the unclosed sutures of infants' heads. This also, I presume, is the reason that the eyes of a strangled man appear as if they were starting from their sockets.

With respect to the protection of the eye during violent expiratory efforts by the pressure of the eyelids, Professor Donders concludes from his various observations that this action certainly limits or entirely removes the dilatation of the vessels.[16] At such times, he adds, we

[16] Prof. Donders remarks (ibid. p. 28), that, "After injury to the eye, after operations, and in some forms of internal inflammation, we attach great value to the uniform support of the closed eyelids, and we increase this in many instances by the application of a bandage. In both cases we carefully endeavour to avoid great expiratory pressure, the disadvantage of which is well known." Mr. Bowman informs me that in the excessive photophobia, accompanying what is called scrofulous ophthalmia in children, when the light is so very painful that during weeks or months it is constantly excluded by the most forcible closure of the lids, he has often been struck on opening the lids by the paleness of the eye, —not an unnatural paleness, but an absence of the redness that might have been expected when the surface is somewhat inflamed, as is then usually the case; and this paleness he is inclined to attribute to the forcible closure of the eyelids.

not unfrequently see the hand involuntarily laid upon the eyelids, as if the better to support and defend the eyeball.

Nevertheless much evidence cannot at present be advanced to prove that the eye actually suffers injury from the want of support during violent expiration; but there is some. It is "a fact that forcible expiratory efforts in violent coughing or vomiting, and especially in sneezing, sometimes give rise to ruptures of the little (external) vessels" of the eye.[17] With respect to the internal vessels, Dr. Gunning has lately recorded a case of exophthalmos in consequence of whooping-cough, which in his opinion depended on the rupture of the deeper vessels; and another analogous case has been recorded. But a mere sense of discomfort would probably suffice to lead to the associated habit of protecting the eyeball by the contraction of the surrounding muscles. Even the expectation or chance of injury would probably be sufficient, in the same manner as an object moving too near the eye induces involuntary winking of the eyelids. We may, therefore, safely conclude from Sir C. Bell's observations, and more especially from the more careful investigations by Professor Donders, that the firm closure of the eyelids during the screaming of children is an action full of meaning and of real service.

We have already seen that the contraction of the orbicular muscles leads to the drawing up of the upper lip, and consequently, if the mouth is kept widely open, to the drawing down of the corners by the contraction of the depressor muscles. The formation of the nasolabial fold on the cheeks likewise follows from the drawing up of the upper lip. Thus all the chief expressive movements of the face during crying apparently result

---

[17] Donders, ibid. p. 36.

from the contraction of the muscles round the eyes. We shall also find that the shedding of tears depends on, or at least stands in some connection with, the contraction of these same muscles.

In some of the foregoing cases, especially in those of sneezing and coughing, it is possible that the contraction of the orbicular muscles may serve in addition to protect the eyes from too severe a jar or vibration. I think so, because dogs and cats, in crunching hard bones, always close their eyelids, and at least sometimes in sneezing; though dogs do not do so whilst barking loudly. Mr. Sutton carefully observed for me a young orang and chimpanzee, and he found that both always closed their eyes in sneezing and coughing, but not whilst screaming violently. I gave a small pinch of snuff to a monkey of the American division, namely, a Cebus, and it closed its eyelids whilst sneezing; but not on a subsequent occasion whilst uttering loud cries.

Cause of the secretion of tears.—It is an important fact which must be considered in any theory of the secretion of tears from the mind being affected, that whenever the muscles round the eyes are strongly and involuntarily contracted in order to compress the blood-vessels and thus to protect the eyes, tears are secreted, often in sufficient abundance to roll down the cheeks. This occurs under the most opposite emotions, and under no emotion at all. The sole exception, and this is only a partial one, to the existence of a relation between the involuntary and strong contraction of these muscles and the secretion of tears is that of young infants, who, whilst screaming violently with their eyelids firmly closed, do not commonly weep until they have attained the age of from two to three or four months. Their eyes, however, become suffused with tears at a much earlier age. It would appear, as already remarked, that the lacrymal glands do not, from the want of practice or some other cause, come to full functional activity at a very early period of life. With children at a somewhat later age, crying out or wailing from any distress is so regularly accompanied by the shedding of tears, that weeping and crying are synonymous terms.[18]

Under the opposite emotion of great joy or amusement, as long as laughter is moderate there is hardly any contraction of the muscles round the eyes, so that there is no frowning; but when peals of loud laughter are uttered, with rapid and violent spasmodic expirations, tears stream down the face. I have more than once noticed the face of a person, after a paroxysm of violent laughter, and I could see that the orbicular muscles and those running to the upper lip were still partially contracted, which together with the tear-stained cheeks gave to the upper half of the face an expression not to be distinguished from that of a child still blubbering from grief. The fact of tears streaming down the face during violent laughter is common to all the races of mankind, as we shall see in a future chapter.

In violent coughing, especially when a person is half-choked, the face becomes purple, the veins distended, the orbicular muscles strongly contracted, and tears run down the cheeks. Even after a fit of ordinary coughing, almost every one has to wipe his eyes. In violent vomiting or retching, as I have myself experienced and seen in others, the orbicular muscles are strongly contracted, and tears sometimes flow freely down the cheeks. It has been suggested to me that this may be due to irritating matter being injected into the nostrils, and caus-

---

[18] Mr. Hensleigh Wedgwood (Dict. of English Etymology, 1859, vol. i. p. 410) says, "the verb to weep comes from Anglo-Saxon *wop*, the primary meaning of which is simply outcry."

ing by reflex action the secretion of tears. Accordingly I asked one of my informants, a surgeon, to attend to the effects of retching when nothing was thrown up from the stomach; and, by an odd coincidence, he himself suffered the next morning from an attack of retching, and three days subsequently observed a lady under a similar attack; and he is certain that in neither case an atom of matter was ejected from the stomach; yet the orbicular muscles were strongly contracted, and tears freely secreted. I can also speak positively to the energetic contraction of these same muscles round the eyes, and to the coincident free secretion of tears, when the abdominal muscles act with unusual force in a downward direction on the intestinal canal.

Yawning commences with a deep inspiration, followed by a long and forcible expiration; and at the same time almost all the muscles of the body are strongly contracted, including those round the eyes. During this act tears are often secreted, and I have seen them even rolling down the cheeks.

I have frequently observed that when persons scratch some point which itches intolerably, they forcibly close their eyelids; but they do not, as I believe, first draw a deep breath and then expel it with force; and I have never noticed that the eyes then become filled with tears; but I am not prepared to assert that this does not occur. The forcible closure of the eyelids is, perhaps, merely a part of that general action by which almost all the muscles of the body are at the same time rendered rigid. It is quite different from the gentle closure of the eyes which often accompanies, as Gratiolet remarks,[19] the smelling a delicious odour, or the tasting a delicious morsel, and which probably originates in the desire to shut out any disturbing impression through the eyes.

[19] 'De la Physionomie,' 1865, p. 217.

Professor Donders writes to me to the following effect: "I have observed some cases of a very curious affection when, after a slight rub (*attouchement*), for example, from the friction of a coat, which caused neither a wound nor a contusion, spasms of the orbicular muscles occurred, with a very profuse flow of tears, lasting about one hour. Subsequently, sometimes after an interval of several weeks, violent spasms of the same muscles re-occurred, accompanied by the secretion of tears, together with primary or secondary redness of the eye." Mr. Bowman informs me that he has occasionally observed closely analogous cases, and that, in some of these, there was no redness or inflammation of the eyes.

I was anxious to ascertain whether there existed in any of the lower animals a similar relation between the contraction of the orbicular muscles during violent expiration and the secretion of tears; but there are very few animals which contract these muscles in a prolonged manner, or which shed tears. The *Macacus maurus*, which formerly wept so copiously in the Zoological Gardens, would have been a fine case for observation; but the two monkeys now there, and which are believed to belong to the same species, do not weep. Nevertheless they were carefully observed by Mr. Bartlett and myself, whilst screaming loudly, and they seemed to contract these muscles; but they moved about their cages so rapidly, that it was difficult to observe with certainty. No other monkey, as far as I have been able to ascertain, contracts its orbicular muscles whilst screaming.

The Indian elephant is known sometimes to weep. Sir E. Tennent, in describing these which he saw captured and bound in Ceylon, says, some "lay motionless on the ground, with no other indication of suffering than the tears which suffused their eyes and flowed incessantly." Speaking of another elephant he says, "When overpowered and made fast, his grief was most affecting; his violence sank to utter prostration, and he lay on the ground, uttering choking cries, with tears trickling down his cheeks."[20] In the Zoological Gardens the keeper of the Indian elephants positively asserts that he has several times seen tears rolling down the face of the old female, when distressed by the removal of the young one. Hence I was extremely anxious to ascertain, as an extension of the relation between the contraction of the orbicular muscles and the shedding of tears in man, whether elephants when screaming or trumpeting loudly contract these muscles. At Mr. Bartlett's desire the keeper ordered the old and the young elephant to trumpet; and we repeatedly saw in both animals that, just as the trumpeting began, the orbicular muscles, especially the lower ones, were distinctly contracted. On a

[20] 'Ceylon,' 3rd edit. 1859, vol. ii. pp. 364, 376. I applied to Mr. Thwaites, in Ceylon, for further information with respect to the weeping of the elephant; and in consequence received a letter from the Rev. Mr Glenie, who, with others, kindly observed for me a herd of recently captured elephants. These, when irritated, screamed violently; but it is remarkable that they never when thus screaming contracted the muscles round the eyes. Nor did they shed tears; and the native hunters asserted that they had never observed elephants weeping. Nevertheless, it appears to me impossible to doubt Sir E. Tennent's distinct details about their weeping, supported as they are by the positive assertion of the keeper in the Zoological Gardens. It is certain that the two elephants in the Gardens, when they began to trumpet loudly, invariably contracted their orbicular muscles. I can reconcile these conflicting statements only by supposing that the recently captured elephants in Ceylon, from being enraged or frightened, desired to observe their persecutors, and consequently did not contract their orbicular muscles, so that their vision might not be impeded. Those seen weeping by Sir E. Tennent were prostrate, and had given up the contest in despair. The elephants which trumpeted in the Zoological Gardens at the word of command, were, of course, neither alarmed nor enraged.

subsequent occasion the keeper made the old elephant trumpet much more loudly, and invariably both the upper and lower orbicular muscles were strongly contracted, and now in an equal degree. It is a singular fact that the African elephant, which, however, is so different from the Indian species that it is placed by some naturalists in a distinct sub-genus, when made on two occasions to trumpet loudly, exhibited no trace of the contraction of the orbicular muscles.

From the several foregoing cases with respect to Man, there can, I think, be no doubt that the contraction of the muscles round the eyes, during violent expiration or when the expanded chest is forcibly compressed, is, in some manner, intimately connected with the secretion of tears. This holds good under widely different emotions, and independently of any emotion. It is not, of course, meant that tears cannot be secreted without the contraction of these muscles; for it is notorious that they are often freely shed with the eyelids not closed, and with the brows unwrinkled. The contraction must be both involuntary and prolonged, as during a choking fit, or energetic, as during a sneeze. The mere involuntary winking of the eyelids, though often repeated, does not bring tears into the eyes. Nor does the voluntary and prolonged contraction of the several surrounding muscles suffice. As the lacrymal glands of children are easily excited, I persuaded my own and several other children of different ages to contract these muscles repeatedly with their utmost force, and to continue doing so as long as they possibly could; but this produced hardly any effect. There was sometimes a little moisture in the eyes, but not more than apparently could be accounted for by the squeezing out of the already secreted tears within the glands.

The nature of the relation between the involuntary

and energetic contraction of the muscles round the eyes, and the secretion of tears, cannot be positively ascertained, but a probable view may be suggested. The primary function of the secretion of tears, together with some mucus, is to lubricate the surface of the eye; and a secondary one, as some believe, is to keep the nostrils damp, so that the inhaled air may be moist,[21] and likewise to favour the power of smelling. But another, and at least equally important function of tears, is to wash out particles of dust or other minute objects which may get into the eyes. That this is of great importance is clear from the cases in which the cornea has been rendered opaque through inflammation, caused by particles of dust not being removed, in consequence of the eye and eyelid becoming immovable.[22] The secretion of tears from the irritation of any foreign body in the eye is a reflex action;—that is, the body irritates a peripheral nerve which sends an impression to certain sensory nerve-cells; these transmit an influence to other cells, and these again to the lacrymal glands. The influence transmitted to these glands causes, as there is good reason to believe, the relaxation of the muscular coats of the smaller arteries; this allows more blood to permeate the glandular tissue, and this induces a free secretion of tears. When the small arteries of the face, including those of the retina, are relaxed under very different circumstances, namely, during an intense blush, the lacrymal glands are sometimes affected in a like manner, for the eyes become suffused with tears.

It is difficult to conjecture how many reflex actions have originated, but, in relation to the present case of

---

[21] Bergeon, as quoted in the 'Journal of Anatomy and Physiology,' Nov. 1871, p. 235.

[22] See, for instance, a case given by Sir Charles Bell, 'Philosophical Transactions,' 1823, p. 177.

the affection of the lacrymal glands through irritation of the surface of the eye, it may be worth remarking that, as soon as some primordial form became semi-terrestrial in its habits, and was liable to get particles of dust into its eyes, if these were not washed out they would cause much irritation; and on the principle of the radiation of nerve-force to adjoining nerve-cells, the lacrymal glands would be stimulated to secretion. As this would often recur, and as nerve-force readily passes along accustomed channels, a slight irritation would ultimately suffice to cause a free secretion of tears.

As soon as by this, or by some other means, a reflex action of this nature had been established and rendered easy, other stimulants applied to the surface of the eye —such as a cold wind, slow inflammatory action, or a blow on the eyelids—would cause a copious secretion of tears, as we know to be the case. The glands are also excited into action through the irritation of adjoining parts. Thus when the nostrils are irritated by pungent vapours, though the eyelids may be kept firmly closed, tears are copiously secreted; and this likewise follows from a blow on the nose, for instance from a boxing-glove. A stinging switch on the face produces, as I have seen, the same effect. In these latter cases the secretion of tears is an incidental result, and of no direct service. As all these parts of the face, including the lacrymal glands, are supplied with branches of the same nerve, namely, the fifth, it is in some degree intelligible that the effects of the excitement of any one branch should spread to the nerve-cells or roots of the other branches.

The internal parts of the eye likewise act, under certain conditions, in a reflex manner on the lacrymal glands. The following statements have been kindly communicated to me by Mr. Bowman; but the subject is a very intricate one, as all the parts of the eye are so intimately related together, and are so sensitive to various stimulants. A strong light acting on the retina, when in a normal condition, has very little tendency to cause lacrymation; but with unhealthy children having small, old-standing ulcers on the cornea, the retina becomes excessively sensitive to light, and exposure even to common daylight causes forcible and sustained closure of the lids, and a profuse flow of tears. When persons who ought to begin the use of convex glasses habitually strain the waning power of accommodation, an undue secretion of tears very often follows, and the retina is liable to become unduly sensitive to light. In general, morbid affections of the surface of the eye, and of the ciliary structures concerned in the accommodative act, are prone to be accompanied with excessive secretion of tears. Hardness of the eyeball, not rising to inflammation, but implying a want of balance between the fluids poured out and again taken up by the intra-ocular vessels, is not usually attended with any lacrymation. When the balance is on the other side, and the eye becomes too soft, there is a greater tendency to lacrymation. Finally, there are numerous morbid states and structural alterations of the eyes, and even terrible inflammations, which may be attended with little or no secretion of tears.

It also deserves notice, as indirectly bearing on our subject, that the eye and adjoining parts are subject to an extraordinary number of reflex and associated movements, sensations, and actions, besides those relating to the lacrymal glands. When a bright light strikes the retina of one eye alone, the iris contracts, but the iris of the other eye moves after a measurable interval of time. The iris likewise moves in accommodation to near or distant vision, and when the two eyes are made to converge.[23] Every one knows how irresistibly the eyebrows are drawn down under an intensely bright light. The eyelids also involuntarily wink when an object is moved near the eyes, or a sound is suddenly heard. The well-known case of a bright light causing some persons to sneeze is even more curious; for nerve-force here radiates from certain nerve-cells in connection with the retina, to the sensory nerve-cells of the nose, causing it to tickle; and from these, to the cells which command the various respiratory muscles (the orbiculars included) which expel the air in so peculiar a manner that it rushes through the nostrils alone.

To return to our point: why are tears secreted during a screaming-fit or other violent expiratory efforts? As a slight blow on the eyelids causes a copious secretion of tears, it is at least possible that the spasmodic contraction of the eyelids, by pressing strongly on the eyeball, should in a similar manner cause some secretion. This seems possible, although the voluntary contraction of the same muscles does not produce any such effect. We know that a man cannot voluntarily sneeze or cough with nearly the same force as he does automatically; and so it is with the contraction of the orbicular muscles: Sir C. Bell experimented on them, and found that by suddenly and forcibly closing the eyelids in the dark, sparks of light are seen, like those caused by tapping the eyelids with the fingers; "but in sneezing the compression is both more rapid and more forcible, and the sparks are more brilliant." That these sparks are due to the contraction of the eyelids is clear, because if they "are held open during the act of sneezing, no sensation of light will be experienced." In the peculiar cases re-

[23] See, on these several points, Prof. Donders 'On the Anomalies of Accommodation and Refraction of the Eye,' 1864, p. 573.

ferred to by Professor Donders and Mr. Bowman, we have seen that some weeks after the eye has been very slightly injured, spasmodic contractions of the eyelids ensue, and these are accompanied by a profuse flow of tears. In the act of yawning, the tears are apparently due solely to the spasmodic contraction of the muscles round the eyes. Notwithstanding these latter cases, it seems hardly credible that the pressure of the eyelids on the surface of the eye, although effected spasmodically and therefore with much greater force than can be done voluntarily, should be sufficient to cause by reflex action the secretion of tears in the many cases in which this occurs during violent expiratory efforts.

Another cause may come conjointly into play. We have seen that the internal parts of the eye, under certain conditions, act in a reflex manner on the lacrymal glands. We know that during violent expiratory efforts the pressure of the arterial blood within the vessels of the eye is increased, and that the return of the venous blood is impeded. It seems, therefore, not improbable that the distension of the ocular vessels, thus induced, might act by reflection on the lacrymal glands—the effects due to the spasmodic pressure of the eyelids on the surface of the eye being thus increased.

In considering how far this view is probable, we should bear in mind that the eyes of infants have been acted on in this double manner during numberless generations, whenever they have screamed; and on the principle of nerve-force readily passing along accustomed channels, even a moderate compression of the eyeballs and a moderate distension of the ocular vessels would ultimately come, through habit, to act on the glands. We have an analogous case in the orbicular muscles being almost always contracted in some slight degree, even during a gentle crying-fit, when there can be no distension of the vessels and no uncomfortable sensation excited within the eyes.

Moreover, when complex actions or movements have long been performed in strict association together, and these are from any cause at first voluntarily and afterwards habitually checked, then if the proper exciting conditions occur, any part of the action or movement which is least under the control of the will, will often still be involuntarily performed. The secretion by a gland is remarkably free from the influence of the will; therefore, when with the advancing age of the individual, or with the advancing culture of the race, the habit of crying out or screaming is restrained, and there is consequently no distension of the blood-vessels of the eye, it may nevertheless well happen that tears should still be secreted. We may see, as lately remarked, the muscles round the eyes of a person who reads a pathetic story, twitching or trembling in so slight a degree as hardly to be detected. In this case there has been no screaming and no distension of the blood-vessels, yet through habit certain nerve-cells send a small amount of nerve-force to the cells commanding the muscles round the eyes; and they likewise send some to the cells commanding the lacrymal glands, for the eyes often become at the same time just moistened with tears. If the twitching of the muscles round the eyes and the secretion of tears had been completely prevented, nevertheless it is almost certain that there would have been some tendency to transmit nerve-force in these same directions; and as the lacrymal glands are remarkably free from the control of the will, they would be eminently liable still to act, thus betraying, though there were no other outward signs, the pathetic thoughts which were passing through the person's mind.

As a further illustration of the view here advanced,

I may remark that if, during an early period of life, when habits of all kinds are readily established, our infants, when pleased, had been accustomed to utter loud peals of laughter (during which the vessels of their eyes are distended) as often and as continuously as they have yielded when distressed to screaming-fits, then it is probable that in after life tears would have been as copiously and as regularly secreted under the one state of mind as under the other. Gentle laughter, or a smile, or even a pleasing thought, would have sufficed to cause a moderate secretion of tears. There does indeed exist an evident tendency in this direction, as will be seen in a future chapter, when we treat of the tender feelings. With the Sandwich Islanders, according to Freycinet,[24] tears are actually recognized as a sign of happiness; but we should require better evidence on this head than that of a passing voyager. So again if our infants, during many generations, and each of them during several years, had almost daily suffered from prolonged choking-fits, during which the vessels of the eye are distended and tears copiously secreted, then it is probable, such is the force of associated habit, that during after life the mere thought of a choke, without any distress of mind, would have sufficed to bring tears into our eyes.

To sum up this chapter, weeping is probably the result of some such chain of events as follows. Children, when wanting food or suffering in any way, cry out loudly, like the young of most other animals, partly as a call to their parents for aid, and partly from any great exertion serving as a relief. Prolonged screaming inevitably leads to the gorging of the blood-vessels of the eye; and this will have led, at first consciously and at

[24] Quoted by Sir J. Lubbock, 'Prehistoric Times,' 1865, p. 458.

last habitually, to the contraction of the muscles round the eyes in order to protect them. At the same time the spasmodic pressure on the surface of the eye, and the distension of the vessels within the eye, without necessarily entailing any conscious sensation, will have affected, through reflex action, the lacrymal glands. Finally, through the three principles of nerve-force readily passing along accustomed channels—of association, which is so widely extended in its power—and of certain actions, being more under the control of the will than others—it has come to pass that suffering readily causes the secretion of tears, without being necessarily accompanied by any other action.

Although in accordance with this view we must look at weeping as an incidental result, as purposeless as the secretion of tears from a blow outside the eye, or as a sneeze from the retina being affected by a bright light, yet this does not present any difficulty in our understanding how the secretion of tears serves as a relief to suffering. And by as much as the weeping is more violent or hysterical, by so much will the relief be greater, —on the same principle that the writhing of the whole body, the grinding of the teeth, and the uttering of piercing shrieks, all give relief under an agony of pain.

CHAPTER VII.

LOW SPIRITS, ANXIETY, GRIEF, DEJECTION, DESPAIR.

General effect of grief on the system—Obliquity of the eyebrows under suffering—On the cause of the obliquity of the eyebrows—On the depression of the corners of the mouth.

AFTER the mind has suffered from an acute paroxysm of grief, and the cause still continues, we fall into a state of low spirits; or we may be utterly cast down and dejected. Prolonged bodily pain, if not amounting to an agony, generally leads to the same state of mind. If we expect to suffer, we are anxious; if we have no hope of relief, we despair.

Persons suffering from excessive grief often seek relief by violent and almost frantic movements, as described in a former chapter; but when their suffering is somewhat mitigated, yet prolonged, they no longer wish for action, but remain motionless and passive, or may occasionally rock themselves to and fro. The circulation becomes languid; the face pale; the muscles flaccid; the eyelids droop; the head hangs on the contracted chest; the lips, cheeks, and lower jaw all sink downwards from their own weight. Hence all the features are lengthened; and the face of a person who hears bad news is said to fall. A party of natives in Tierra del Fuego endeavoured to explain to us that their friend, the captain of a sealing vessel, was out of spirits, by pulling down their cheeks with both hands, so as to make their faces as long as possible. Mr. Bunnet informs me that the Australian aborigines when out of spirits have a chop-fallen appearance. After prolonged suffering the eyes become dull and lack expression, and are often slightly suffused with tears. The eyebrows not rarely are rendered oblique, which is due to their inner ends being raised. This produces peculiarly-formed wrinkles on the forehead, which are very different from those of a simple frown; though in some cases a frown alone may be present. The corners of the mouth are drawn downwards, which is so universally recognized as a sign of being out of spirits, that it is almost proverbial.

The breathing becomes slow and feeble, and is often interrupted by deep sighs. As Gratiolet remarks, whenever our attention is long concentrated on any subject, we forget to breathe, and then relieve ourselves by a deep inspiration; but the sighs of a sorrowful person, owing to his slow respiration and languid circulation, are eminently characteristic.[1] As the grief of a person in this state occasionally recurs and increases into a paroxysm, spasms affect the respiratory muscles, and he feels as if something, the so-called *globus hystericus*, was rising in his throat. These spasmodic movements are clearly allied to the sobbing of children, and are remnants of those severer spasms which occur when a person is said to choke from excessive grief.[2]

---

[1] The above descriptive remarks are taken in part from my own observations, but chiefly from Gratiolet ('De la Physionomie,' pp. 53, 337; on Sighing, 232), who has well treated this whole subject. See, also, Huschke, 'Mimices et Physiognomices, Fragmentum Physiologicum,' 1821, p. 21. On the dulness of the eyes, Dr. Piderit, 'Mimik und Physiognomik,' 1867, s. 65.

[2] On the action of grief on the organs of respiration,

Obliquity of the eyebrows.—Two points alone in the above description require further elucidation, and these are very curious ones; namely, the raising of the inner ends of the eyebrows, and the drawing down of the corners of the mouth. With respect to the eyebrows, they may occasionally be seen to assume an oblique position in persons suffering from deep dejection or anxiety; for instance, I have observed this movement in a mother whilst speaking about her sick son; and it is sometimes excited by quite trifling or momentary causes of real or pretended distress. The eyebrows assume this position owing to the contraction of certain muscles (namely, the orbiculars, corrugators, and pyramidals of the nose, which together tend to lower and contract the eyebrows) being partially checked by the more powerful action of the central fasciæ of the frontal muscle. These latter fasciæ by their contraction raise the inner ends alone of the eyebrows; and as the corrugators at the same time draw the eyebrows together, their inner ends become puckered into a fold or lump. This fold is a highly characteristic point in the appearance of the eyebrows when rendered oblique, as may be seen in figs. 2 and 5, Plate II. The eyebrows are at the same time somewhat roughened, owing to the hairs being made to project. Dr. J. Crichton Browne has also often noticed in melancholic patients who keep their eyebrows persistently oblique, "a peculiar acute arching of the upper eyelid." A trace of this may be observed by comparing the right and left eyelids of the young man in the photograph (fig. 2, Plate II.); for he was not able to act equally on both eyebrows. This is also shown by the unequal furrows on the two sides of his forehead. The acute arching of the eyelids

---

see more especially Sir C. Bell, 'Anatomy of Expression,' 3rd edit. 1844, p. 151.

PLATE II

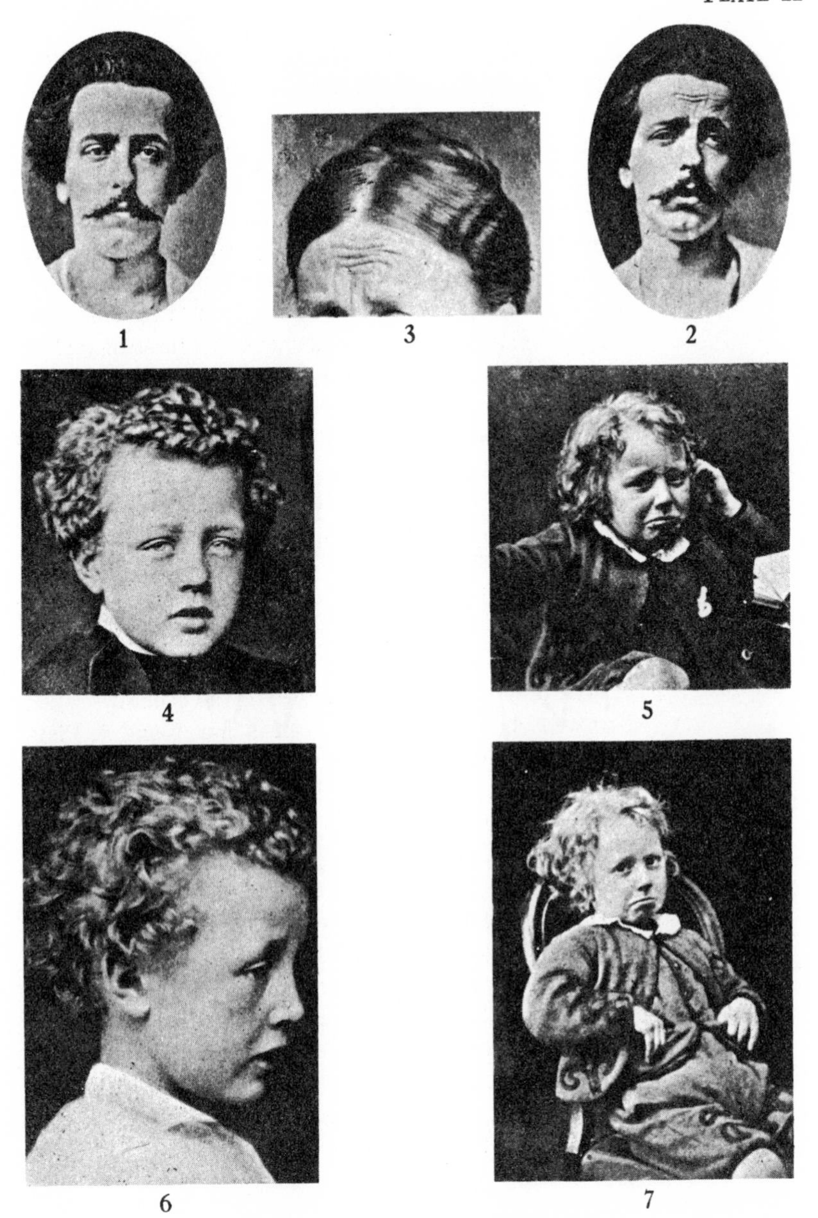

depends, I believe, on the inner end alone of the eyebrows being raised; for when the whole eyebrow is elevated and arched, the upper eyelid follows in a slight degree the same movement.

But the most conspicuous result of the opposed contraction of the above-named muscles, is exhibited by the peculiar furrows formed on the forehead. These muscles, when thus in conjoint yet opposed action, may be called, for the sake of brevity, the grief-muscles. When a person elevates his eyebrows by the contraction of the whole frontal muscle, transverse wrinkles extend across the whole breadth of the forehead; but in the present case the middle fasciæ alone are contracted; consequently, transverse furrows are formed across the middle part alone of the forehead. The skin over the exterior parts of both eyebrows is at the same time drawn downwards and smooth, by the contraction of the outer portions of the orbicular muscles. The eyebrows are likewise brought together through the simultaneous contraction of the corrugators;[3] and this latter action generates

[3] In the foregoing remarks on the manner in which the eyebrows are made oblique, I have followed what seems to be the universal opinion of all the anatomists, whose works I have consulted on the action of the above-named muscles, or with whom I have conversed. Hence throughout this work I shall take a similar view of the action of the *corrugator supercilii*, *orbicularis*, *pyramidalis nasi*, and *frontalis* muscles. Dr. Duchenne, however, believes, and every conclusion at which he arrives deserves serious consideration, that it is the corrugator, called by him the *sourcilier*, which raises the inner corner of the eyebrows and is antagonistic to the upper and inner part of the orbicular muscle, as well as to the *pyramidalis nasi* (see Mécanisme de la Phys. Humaine, 1862, folio, art. v., text and figures 19 to 29: octavo edit. 1862, p. 43 text). He admits, however, that the corrugator draws together the eyebrows, causing vertical furrows above the base of the nose, or a frown. He further believes that towards the outer two-thirds of the eyebrow the corrugator acts in conjunction with the upper orbicular muscle; both here standing in antagonism to the frontal muscle. I

vertical furrows, separating the exterior and lowered part of the skin of the forehead from the central and raised part. The union of these vertical furrows with the central and transverse furrows (see figs. 2 and 3) produces a mark on the forehead which has been compared to a horse-shoe; but the furrows more strictly form three sides of a quadrangle. They are often conspicuous on the foreheads of adult or nearly adult persons, when their eyebrows are made oblique; but with young children, owing to their skin not easily wrinkling, they are rarely seen, or mere traces of them can be detected.

These peculiar furrows are best represented in fig. 3, Plate II., on the forehead of a young lady who has the power in an unusual degree of voluntarily acting on the requisite muscles. As she was absorbed in the attempt, whilst being photographed, her expression was not at all one of grief; I have therefore given the forehead alone. Fig. 1 on the same plate, copied from Dr. Duchenne's work,[4] represents, on a reduced scale, the face, in its natural state, of a young man who was a good actor. In fig. 2 he is shown simulating grief, but the

---

am unable to understand, judging from Henle's drawings (woodcut, fig. 3), how the corrugator can act in the manner described by Duchenne. See, also, on this subject, Prof. Donders' remarks in the 'Archives of Medicine,' 1870, vol. v. p. 34. Mr. J. Wood, who is so well known for his careful study of the muscles of the human frame, informs me that he believes the account which I have given of the action of the corrugator to be correct. But this is not a point of any importance with respect to the expression which is caused by the obliquity of the eyebrows, nor of much importance to the theory of its origin.

[4] I am greatly indebted to Dr. Duchenne for permission to have these two photographs (figs. 1 and 2) reproduced by the heliotype process from his work in folio. Many of the foregoing remarks on the furrowing of the skin, when the eyebrows are rendered oblique, are taken from his excellent discussion on this subject.

two eyebrows, as before remarked, are not equally acted on. That the expression is true, may be inferred from the fact that out of fifteen persons, to whom the original photograph was shown, without any clue to what was intended being given them, fourteen immediately answered, "despairing sorrow," "suffering endurance," "melancholy," and so forth. The history of fig. 5 is rather curious: I saw the photograph in a shop-window, and took it to Mr. Rejlander for the sake of finding out by whom it had been made; remarking to him how pathetic the expression was. He answered, "I made it, and it was likely to be pathetic, for the boy in a few minutes burst out crying." He then showed me a photograph of the same boy in a placid state, which I have had (fig. 4) reproduced. In fig. 6, a trace of obliquity in the eyebrows may be detected; but this figure, as well as fig. 7, is given to show the depression of the corners of the mouth, to which subject I shall presently refer.

Few persons, without some practice, can voluntarily act on their grief-muscles; but after repeated trials a considerable number succeed, whilst others never can. The degree of obliquity in the eyebrows, whether assumed voluntarily or unconsciously, differs much in different persons. With some who apparently have unusually strong pyramidal muscles, the contraction of the central fasciæ of the frontal muscle, although it may be energetic, as shown by the quadrangular furrows on the forehead, does not raise the inner ends of the eyebrows, but only prevents their being so much lowered as they otherwise would have been. As far as I have been able to observe, the grief-muscles are brought into action much more frequently by children and women than by men. They are rarely acted on, at least with grown-up persons, from bodily pain, but almost exclusively from mental distress. Two persons who, after some practice, succeeded in acting on their grief-muscles, found by looking at a mirror that when they made their eyebrows oblique, they unintentionally at the same time depressed the corners of their mouths; and this is often the case when the expression is naturally assumed.

The power to bring the grief-muscles freely into play appears to be hereditary, like almost every other human faculty. A lady belonging to a family famous for having produced an extraordinary number of great actors and actresses, and who can herself give this expression "with singular precision," told Dr. Crichton Browne that all her family had possessed the power in a remarkable degree. The same hereditary tendency is said to have extended, as I likewise hear from Dr. Browne, to the last descendant of the family, which gave rise to Sir Walter Scott's novel of 'Red Gauntlet;' but the hero is described as contracting his forehead into a horseshoe mark from any strong emotion. I have also seen a young woman whose forehead seemed almost habitually thus contracted, independently of any emotion being at the time felt.

The grief-muscles are not very frequently brought into play; and as the action is often momentary, it easily escapes observation. Although the expression, when observed, is universally and instantly recognized as that of grief or anxiety, yet not one person out of a thousand who has never studied the subject, is able to say precisely what change passes over the sufferer's face. Hence probably it is that this expression is not even alluded to, as far as I have noticed, in any work of fiction, with the exception of 'Red Gauntlet' and of one other novel; and the authoress of the latter, as I am informed, belongs to the famous family of actors just alluded to; so that her attention may have been specially called to the subject.

The ancient Greek sculptors were familiar with the expression, as shown in the statues of the Laocoon and Arretino; but, as Duchenne remarks, they carried the transverse furrows across the whole breadth of the forehead, and thus committed a great anatomical mistake: this is likewise the case in some modern statues. It is, however, more probable that these wonderfully accurate observers intentionally sacrificed truth for the sake of beauty, than that they made a mistake; for rectangular furrows on the forehead would not have had a grand appearance on the marble. The expression, in its fully developed condition, is, as far as I can discover, not often represented in pictures by the old masters, no doubt owing to the same cause; but a lady who is perfectly familiar with this expression, informs me that in Fra Angelico's 'Descent from the Cross,' in Florence, it is clearly exhibited in one of the figures on the right-hand; and I could add a few other instances.

Dr. Crichton Browne, at my request, closely attended to this expression in the numerous insane patients under his care in the West Riding Asylum; and he is familiar with Duchenne's photographs of the action of the grief-muscles. He informs me that they may constantly be seen in energetic action in cases of melancholia, and especially of hypochondria; and that the persistent lines or furrows, due to their habitual contraction, are characteristic of the physiognomy of the insane belonging to these two classes. Dr. Browne carefully observed for me during a considerable period three cases of hypochondria, in which the grief-muscles were persistently contracted. In one of these, a widow, aged 51, fancied that she had lost all her viscera, and that her whole body was empty. She wore an expression of great distress, and beat her semi-closed hands rhythmically together for hours. The grief-muscles were permanently contracted, and the upper eyelids arched. This condition lasted for months; she then recovered, and her countenance resumed its natural expression. A second case presented nearly the same peculiarities, with the addition that the corners of the mouth were depressed.

Mr. Patrick Nicol has also kindly observed for me several cases in the Sussex Lunatic Asylum, and has communicated to me full details with respect to three of them; but they need not here be given. From his observations on melancholic patients, Mr. Nicol concludes that the inner ends of the eyebrows are almost always more or less raised, with the wrinkles on the forehead more or less plainly marked. In the case of one young woman, these wrinkles were observed to be in constant slight play or movement. In some cases the corners of the mouth are depressed, but often only in a slight degree. Some amount of difference in the expression of the several melancholic patients could almost always be observed. The eyelids generally droop; and the skin near their outer corners and beneath them is wrinkled. The naso-labial fold, which runs from the wings of the nostrils to the corners of the mouth, and which is so conspicuous in blubbering children, is often plainly marked in these patients.

Although with the insane the grief-muscles often act persistently; yet in ordinary cases they are sometimes brought unconsciously into momentary action by ludicrously slight causes. A gentleman rewarded a young lady by an absurdly small present; she pretended to be offended, and as she upbraided him, her eyebrows became extremely oblique, with the forehead properly wrinkled. Another young lady and a youth, both in the highest spirits, were eagerly talking together with extraordinary rapidity; and I noticed that, as often as the young lady was beaten, and could not get out her words fast enough, her eyebrows went obliquely upwards, and rectangular furrows were formed on her forehead. She thus each time hoisted a flag of distress; and this she did half-a-dozen times in the course of a few minutes. I made no remark on the subject, but on a subsequent occasion I asked her to act on her grief-muscles; another girl who was present, and who could do so voluntarily, showing her what was intended. She tried repeatedly, but utterly failed; yet so slight a cause of distress as not being able to talk quickly enough, sufficed to bring these muscles over and over again into energetic action.

The expression of grief, due to the contraction of the grief-muscles, is by no means confined to Europeans, but appears to be common to all the races of mankind. I have, at least, received trustworthy accounts in regard to Hindoos, Dhangars (one of the aboriginal hill-tribes of India, and therefore belonging to a quite distinct race from the Hindoos), Malays, Negroes and Australians. With respect to the latter, two observers answer my query in the affirmative, but enter into no details. Mr. Taplin, however, appends to my descriptive remarks the words "this is exact." With respect to negroes, the lady who told me of Fra Angelico's picture, saw a negro towing a boat on the Nile, and as he encountered an obstruction, she observed his grief-muscles in strong action, with the middle of the forehead well wrinkled. Mr. Geach watched a Malay man in Malacca, with the corners of his mouth much depressed, the eyebrows oblique, with deep short grooves on the forehead. This expression lasted for a very short time; and Mr. Geach remarks it "was a strange one, very much like a person about to cry at some great loss."

In India Mr. H. Erskine found that the natives were familiar with this expression; and Mr. J. Scott, of the

Botanic Gardens, Calcutta, has obligingly sent me a full description of two cases. He observed during some time, himself unseen, a very young Dhangar woman from Nagpore, the wife of one of the gardeners, nursing her baby who was at the point of death; and he distinctly saw the eyebrows raised at the inner corners, the eyelids drooping, the forehead wrinkled in the middle, the mouth slightly open, with the corners much depressed. He then came from behind a screen of plants and spoke to the poor woman, who started, burst into a bitter flood of tears, and besought him to cure her baby. The second case was that of a Hindustani man, who from illness and poverty was compelled to sell his favourite goat. After receiving the money, he repeatedly looked at the money in his hand and then at the goat, as if doubting whether he would not return it. He went to the goat, which was tied up ready to be led away, and the animal reared up and licked his hands. His eyes then wavered from side to side; his "mouth was partially closed, with the corners very decidedly depressed." At last the poor man seemed to make up his mind that he must part with his goat, and then, as Mr. Scott saw, the eyebrows became slightly oblique, with the characteristic puckering or swelling at the inner ends, but the wrinkles on the forehead were not present. The man stood thus for a minute, then heaving a deep sigh, burst into tears, raised up his two hands, blessed the goat, turned round, and without looking again, went away.

On the cause of the obliquity of the eyebrows under suffering.—During several years no expression seemed to me so utterly perplexing as this which we are here considering. Why should grief or anxiety cause the central fasciæ alone of the frontal muscle together with those round the eyes, to contract? Here we seem to have a complex movement for the sole purpose of expressing grief; and yet it is a comparatively rare expression, and often overlooked. I believe the explanation is not so difficult as it at first appears. Dr. Duchenne gives a photograph of the young man before referred to, who, when looking upwards at a strongly illuminated surface, involuntarily contracted his grief-muscles in an exaggerated manner. I had entirely forgotten this photograph, when on a very bright day with the sun behind me, I met, whilst on horseback, a girl whose eyebrows, as she looked up at me, became extremely oblique, with the proper furrows on her forehead. I have observed the same movement under similar circumstances on several subsequent occasions. On my return home I made three of my children, without giving them any clue to my object, look as long and as attentively as they could, at the summit of a tall tree standing against an extremely bright sky. With all three, the orbicular, corrugator, and pyramidal muscles were energetically contracted, through reflex action, from the excitement of the retina, so that their eyes might be protected from the bright light. But they tried their utmost to look upwards; and now a curious struggle, with spasmodic twitchings, could be observed between the whole or only the central portion of the frontal muscle, and the several muscles which serve to lower the eyebrows and close the eyelids. The involuntary contraction of the pyramidal caused the basal part of their noses to be transversely and deeply wrinkled. In one of the three children, the whole eyebrows were momentarily raised and lowered by the alternate contraction of the whole frontal muscle and of the muscles surrounding the eyes, so that the whole breadth of the forehead was alternately wrinkled and smoothed. In the other two children the forehead became wrinkled in the middle part alone, rectangular furrows being thus produced; and the eyebrows were rendered oblique, with their inner extremities puckered and swollen;—in the one child in a slight degree, in the other in a strongly marked manner. This difference in the obliquity of the eyebrows apparently depended on a difference in their general mobility, and in the strength of the pyramidal muscles. In both these cases the eyebrows and forehead were acted on under the influence of a strong light, in precisely the same manner, in every characteristic detail, as under the influence of grief or anxiety.

Duchenne states that the pyramidal muscle of the nose is less under the control of the will than are the other muscles round the eyes. He remarks that the young man who could so well act on his grief-muscles, as well as on most of his other facial muscles, could not contract the pyramidals.[5] This power, however, no doubt differs in different persons. The pyramidal muscle serves to draw down the skin of the forehead between the eyebrows, together with their inner extremities. The central fasciæ of the frontal are the antagonists of the pyramidal; and if the action of the latter is to be specially checked, these central fasciæ must be contracted. So that with persons having powerful pyramidal muscles, if there is under the influence of a bright light an unconscious desire to prevent the lowering of the eyebrows, the central fasciæ of the frontal muscle must be brought into play; and their contraction, if sufficiently strong to overmaster the pyramidals, together with the contraction of the corrugator and orbicular muscles, will act in the manner just described on the eyebrows and forehead.

When children scream or cry out, they contract, as we know, the orbicular, corrugator, and pyramidal mus-

[5] Mécanisme de la Phys. Humaine, Album, p. 15.

cles, primarily for the sake of compressing their eyes, and thus protecting them from being gorged with blood, and secondarily through habit. I therefore expected to find with children, that when they endeavoured either to prevent a crying-fit from coming on, or to stop crying, they would check the contraction of the above-named muscles, in the same manner as when looking upwards at a bright light; and consequently that the central fasciæ of the frontal muscle would often be brought into play. Accordingly, I began myself to observe children at such times, and asked others, including some medical men, to do the same. It is necessary to observe carefully, as the peculiar opposed action of these muscles is not nearly so plain in children, owing to their foreheads not easily wrinkling, as in adults. But I soon found that the grief-muscles were very frequently brought into distinct action on these occasions. It would be superfluous to give all the cases which have been observed; and I will specify only a few. A little girl, a year and a half old, was teased by some other children, and before bursting into tears her eyebrows became decidedly oblique. With an older girl the same obliquity was observed, with the inner ends of the eyebrows plainly puckered; and at the same time the corners of the mouth were drawn downwards. As soon as she burst into tears, the features all changed and this peculiar expression vanished. Again, after a little boy had been vaccinated, which made him scream and cry violently, the surgeon gave him an orange brought for the purpose, and this pleased the child much; as he stopped crying all the characteristic movements were observed, including the formation of rectangular wrinkles in the middle of the forehead. Lastly, I met on the road a little girl three or four years old, who had been frightened by a dog, and when I asked her what was the matter, she stopped whimpering, and her eyebrows instantly became oblique to an extraordinary degree.

Here then, as I cannot doubt, we have the key to the problem why the central fasciæ of the frontal muscle and the muscles round the eyes contract in opposition to each other under the influence of grief;—whether their contraction be prolonged, as with the melancholic insane, or momentary, from some trifling cause of distress. We have all of us, as infants, repeatedly contracted our orbicular, corrugator, and pyramidal muscles, in order to protect our eyes whilst screaming; our progenitors before us have done the same during many generations; and though with advancing years we easily prevent, when feeling distressed, the utterance of screams, we cannot from long habit always prevent a slight contraction of the above-named muscles; nor indeed do we observe their contraction in ourselves, or attempt to stop it, if slight. But the pyramidal muscles seem to be less under the command of the will than the other related muscles; and if they be well developed, their contraction can be checked only by the antagonistic contraction of the central fasciæ of the frontal muscle. The result which necessarily follows, if these fasciæ contract energetically, is the oblique drawing up of the eyebrows, the puckering of their inner ends, and the formation of rectangular furrows on the middle of the forehead. As children and women cry much more freely than men, and as grown-up persons of both sexes rarely weep except from mental distress, we can understand why the grief-muscles are more frequently seen in action, as I believe to be the case, with children and women than with men; and with adults of both sexes from mental distress alone. In some of the cases before recorded, as in that of the poor Dhangar woman and of the Hindustani man, the action of the grief-muscles was quickly followed by bitter weeping. In all cases of distress, whether great or small, our brains tend through long habit to send an order to certain muscles to contract, as if we were still infants on the point of screaming out; but this order we, by the wondrous power of the will, and through habit, are able partially to counteract; although this is effected unconsciously, as far as the means of counteraction are concerned.

On the depression of the corners of the mouth.—This action is effected by the *depressores anguili oris* (see letter K in figs. 1 and 2). The fibres of this muscle diverge downwards, with the upper convergent ends attached round the angles of the mouth, and to the lower lip a little way within the angles.[6] Some of the fibres appear to be antagonistic to the great zygomatic muscle, and others to the several muscles running to the outer part of the upper lip. The contraction of this muscle draws downwards and outwards the corners of the mouth, including the outer part of the upper lip, and even in a slight degree the wings of the nostrils. When the mouth is closed and this muscle acts, the commissure or line of junction of the two lips forms a curved line with the concavity downwards,[7] and the lips themselves are generally somewhat protruded, especially the lower one. The mouth in this state is well represented in the two photographs (Plate II., figs. 6 and 7) by Mr. Rejlander. The upper boy (fig. 6) had just stopped crying, after receiving a slap on the face from another boy; and the right moment was seized for photographing him.

---

[6] Henle, Handbuch der Anat. des Menschen, 1858, B. i. s. 148, figs. 68 and 69.

[7] See the account of the action of this muscle by Dr. Duchenne, 'Mécanisme de la Physionomie Humaine, Album (1862), viii. p. 34.

The expression of low spirits, grief or dejection, due to the contraction of this muscle has been noticed by every one who has written on the subject. To say that a person "is down in the mouth," is synonymous with saying that he is out of spirits. The depression of the corners may often be seen, as already stated on the authority of Dr. Crichton Browne and Mr. Nicol, with the melancholic insane, and was well exhibited in some photographs sent to me by the former gentleman, of patients with a strong tendency to suicide. It has been observed with men belonging to various races, namely with Hindoos, the dark hill-tribes of India, Malays, and, as the Rev. Mr. Hagenauer informs me, with the aborigines of Australia.

When infants scream they firmly contract the muscles round their eyes, and this draws up the upper lip; and as they have to keep their mouths widely open, the depressor muscles running to the corners are likewise brought into strong action. This generally, but not invariably, causes a slight angular bend in the lower lip on both sides, near the corners of the mouth. The result of the upper and lower lip being thus acted on, is that the mouth assumes a squarish outline. The contraction of the depressor muscle is best seen in infants when not screaming violently, and especially just before they begin, or when they cease to scream. Their little faces then acquire an extremely piteous expression, as I continually observed with my own infants between the ages of about six weeks and two or three months. Sometimes, when they are struggling against a crying-fit, the outline of the mouth is curved in so exaggerated a manner as to be like a horseshoe; and the expression of misery then becomes a ludicrous caricature.

The explanation of the contraction of this muscle, under the influence of low spirits or dejection, apparently follows from the same general principles as in the case of the obliquity of the eyebrows. Dr. Duchenne informs me that he concludes from his observations, now prolonged during many years, that this is one of the facial muscles which is least under the control of the will. This fact may indeed be inferred from what has just been stated with respect to infants when doubtfully beginning to cry, or endeavouring to stop crying; for they then generally command all the other facial muscles more effectually than they do the depressors of the corners of the mouth. Two excellent observers who had no theory on the subject, one of them a surgeon, carefully watched for me some older children and women as with some opposed struggling they very gradually approached the point of bursting out into tears; and both observers felt sure that the depressors began to act before any of the other muscles. Now as the depressors have been repeatedly brought into strong action during infancy in many generations, nerve-force will tend to flow, on the principle of long associated habit, to these muscles as well as to various other facial muscles, whenever in after life even a slight feeling of distress is experienced. But as the depressors are somewhat less under the control of the will than most of the other muscles, we might expect that they would often slightly contract, whilst the others remained passive. It is remarkable how small a depression of the corners of the mouth gives to the countenance an expression of low spirits or dejection, so that an extremely slight contraction of these muscles would be sufficient to betray this state of mind.

I may here mention a trifling observation, as it will serve to sum up our present subject. An old lady with a comfortable but absorbed expression sat nearly opposite to me in a railway carriage. Whilst I was looking at her, I saw that her *depressores anguli oris* became very slightly, yet decidedly, contracted; but as her countenance remained as placid as ever, I reflected how meaningless was this contraction, and how easily one might be deceived. The thought had hardly occurred to me when I saw that her eyes suddenly became suffused with tears almost to overflowing, and her whole countenance fell. There could now be no doubt that some painful recollection, perhaps that of a long-lost child, was passing through her mind. As soon as her sensorium was thus affected, certain nerve-cells from long habit instantly transmitted an order to all the respiratory muscles, and to those round the mouth, to prepare for a fit of crying. But the order was countermanded by the will, or rather by a later acquired habit, and all the muscles were obedient, excepting in a slight degree the *depressores anguli oris*. The mouth was not even opened; the respiration was not hurried; and no muscle was affected except those which draw down the corners of the mouth.

As soon as the mouth of this lady began, involuntarily and unconsciously on her part, to assume the proper form for a crying-fit, we may feel almost sure that some nerve-influence would have been transmitted through the long accustomed channels to the various respiratory muscles, as well as to those round the eyes, and to the vaso-motor centre which governs the supply of blood sent to the lacrymal glands. Of this latter fact we have indeed clear evidence in her eyes becoming slightly suffused with tears; and we can understand this, as the lacrymal glands are less under the control of the will than the facial muscles. No doubt there existed at the same time some tendency in the muscles round the eyes at contract, as if for the sake of protecting them from being gorged with blood, but this contraction was completely overmastered, and her brow remained unruffled. Had the pyramidal, corrugator, and orbicular muscles been as little obedient to the will, as they are in many persons, they would have been slightly acted on; and then the central fasciæ of the frontal muscle would have contracted in antagonism, and her eyebrows would have become oblique, with rectangular furrows on her forehead. Her countenance would then have expressed still more plainly than it did a state of dejection, or rather one of grief.

Through steps such as these we can understand how it is, that as soon as some melancholy thought passes through the brain, there occurs a just perceptible drawing down of the corners of the mouth, or a slight raising up of the inner ends of the eyebrows, or both movements combined, and immediately afterwards a slight suffusion of tears. A thrill of nerve-force is transmitted along several habitual channels, and produces an effect on any point where the will has not acquired through long habit much power of interference. The above actions may be considered as rudimental vestiges of the screaming-fits, which are so frequent and prolonged during infancy. In this case, as well as in many others, the links are indeed wonderful which connect cause and effect in giving rise to various expressions on the human countenance; and they explain to us the meaning of certain movements, which we involuntarily and unconsciously perform, whenever certain transitory emotions pass through our minds.

CHAPTER VIII.

JOY, HIGH SPIRITS, LOVE, TENDER FEELINGS, DEVOTION.

Laughter primarily the expression of joy—Ludicrous ideas—Movements of the features during laughter—Nature of the sound produced—The secretion of tears during loud laughter—Gradation from loud laughter to gentle smiling—High spirits—The expression of love—Tender feelings—Devotion.

JOY, when intense, leads to various purposeless movements—to dancing about, clapping the hands, stamping, &c., and to loud laughter. Laughter seems primarily to be the expression of mere joy or happiness. We clearly see this in children at play, who are almost incessantly laughing. With young persons past childhood, when they are in high spirits, there is always much meaningless laughter. The laughter of the gods is described by Homer as "the exuberance of their celestial joy after their daily banquet." A man smiles—and smiling, as we shall see, graduates into laughter—at meeting an old friend in the street, as he does at any trifling pleasure, such as smelling a sweet perfume.[1] Laura Bridgman, from her blindness and deafness, could not have acquired any expression through imitation, yet when a letter from a beloved friend was communicated to her by gesture-language, she "laughed and

[1] Herbert Spencer, 'Essays Scientific,' &c., 1858, p. 360.

clapped her hands, and the colour mounted to her cheeks." On other occasions she has been seen to stamp for joy.[2]

Idiots and imbecile persons likewise afford good evidence that laughter or smiling primarily expresses mere happiness or joy. Dr. Crichton Browne, to whom, as on so many other occasions, I am indebted for the results of his wide experience, informs me that with idiots laughter is the most prevalent and frequent of all the emotional expressions. Many idiots are morose, passionate, restless, in a painful state of mind, or utterly stolid, and these never laugh. Others frequently laugh in a quite senseless manner. Thus an idiot boy, incapable of speech, complained to Dr. Browne, by the aid of signs, that another boy in the asylum had given him a black eye; and this was accompanied by "explosions of laughter and with his face covered with the broadest smiles." There is another large class of idiots who are persistently joyous and benign, and who are constantly laughing or smiling.[3] Their countenances often exhibit a stereotyped smile; their joyousness is increased, and they grin, chuckle, or giggle, whenever food is placed before them, or when they are caressed, are shown bright colours, or hear music. Some of them laugh more than usual when they walk about, or attempt any muscular exertion. The joyousness of most of these idiots cannot possibly be associated, as Dr. Browne remarks, with any distinct ideas: they simply feel pleasure, and express it by laughter or smiles. With imbeciles rather higher in the scale, personal vanity seems to be the commonest cause of laughter, and next to this, pleasure arising from the approbation of their conduct.

---

[2] F. Lieber on the vocal sounds of L. Bridgman, 'Smithsonian Contributions,' 1851, vol. ii. p. 6.

[3] See, also, Mr. Marshall, in Phil. Transact. 1864, p. 526.

With grown-up persons laughter is excited by causes considerably different from those which suffice during childhood; but this remark hardly applies to smiling. Laughter in this respect is analogous with weeping, which with adults is almost confined to mental distress, whilst with children it is excited by bodily pain or any suffering, as well as by fear or rage. Many curious discussions have been written on the causes of laughter with grown-up persons. The subject is extremely complex. Something incongruous or unaccountable, exciting surprise and some sense of superiority in the laugher, who must be in a happy frame of mind, seems to be the commonest cause.[4] The circumstances must not be of a momentous nature: no poor man would laugh or smile on suddenly hearing that a large fortune had been bequeathed to him. If the mind is strongly excited by pleasurable feelings, and any little unexpected event or thought occurs, then, as Mr. Herbert Spencer remarks,[5] "a large amount of nervous energy, instead of being allowed to expend itself in producing an equivalent amount of the new thoughts and emotion which were nascent, is suddenly checked in its flow." . . . "The excess must discharge itself in some other direction, and there results an efflux through the motor nerves to various classes of the muscles, producing the half-convulsive actions we term laughter." An observation, bearing on this point, was made by a correspondent during the recent siege of Paris, namely, that the German soldiers, after strong excitement from exposure to extreme

[4] Mr. Bain ('The Emotions and the Will,' 1865, p. 247) has a long and interesting discussion on the Ludicrous. The quotation above given about the laughter of the gods is taken from this work. See, also, Mandeville, 'The Fable of the Bees,' vol. ii. p. 168.

[5] 'The Physiology of Laughter,' Essays, Second Series, 1863, p. 114.

danger, were particularly apt to burst out into loud laughter at the smallest joke. So again when young children are just beginning to cry, an unexpected event will sometimes suddenly turn their crying into laughter, which apparently serves equally well to expend their superfluous nervous energy.

The imagination is sometimes said to be tickled by a ludicrous idea; and this so-called tickling of the mind is curiously analogous with that of the body. Every one knows how immoderately children laugh, and how their whole bodies are convulsed when they are tickled. The anthropoid apes, as we have seen, likewise utter a reiterated sound, corresponding with our laughter, when they are tickled, especially under the armpits. I touched with a bit of paper the sole of the foot of one of my infants, when only seven days old, and it was suddenly jerked away and the toes curled about, as in an older child. Such movements, as well as laughter from being tickled, are manifestly reflex actions; and this is likewise shown by the minute unstriped muscles, which serve to erect the separate hairs on the body, contracting near a tickled surface.[6] Yet laughter from a ludicrous idea, though involuntary, cannot be called a strictly reflex action. In this case, and in that of laughter from being tickled, the mind must be in a pleasurable condition; a young child, if tickled by a strange man, would scream from fear. The touch must be light, and an idea or event, to be ludicrous, must not be of grave import. The parts of the body which are most easily tickled are those which are not commonly touched, such as the armpits or between the toes, or parts such as the soles of the feet, which are habitually touched by a broad

[6] J. Lister in 'Quarterly Journal of Microscopical Science,' 1853, vol. i. p. 266.

surface; but the surface on which we sit offers a marked exception to this rule. According to Gratiolet,[7] certain nerves are much more sensitive to tickling than others. From the fact that a child can hardly tickle itself, or in a much less degree than when tickled by another person, it seems that the precise point to be touched must not be known; so with the mind, something unexpected—a novel or incongruous idea which breaks through an habitual train of thought—appears to be a strong element in the ludicrous.

The sound of laughter is produced by a deep inspiration followed by short, interrupted, spasmodic contractions of the chest, and especially of the diaphragm.[8] Hence we hear of "laughter holding both his sides." From the shaking of the body, the head nods to and fro. The lower jaw often quivers up and down, as is likewise the case with some species of baboons, when they are much pleased.

During laughter the mouth is opened more or less widely, with the corners drawn much backwards, as well as a little upwards; and the upper lip is somewhat raised. The drawing back of the corners is best seen in moderate laughter, and especially in a broad smile—the latter epithet showing how the mouth is widened. In the accompanying figs. 1–3, Plate III., different degrees of moderate laughter and smiling have been photographed. The figure of the little girl, with the hat, is by Dr. Wallich, and the expression was a genuine one; the other two are by Mr. Rejlander. Dr. Duchenne repeatedly insists[9] that, under the emotion of joy, the

[7] 'De la Physionomie,' p. 186.

[8] Sir C. Bell (Anat. of Expression, p. 147) makes some remarks on the movement of the diaphragm during laughter.

[9] 'Mécanisme de la Physionomie Humaine,' Album, Légende vi.

1 4

2 5

3 6

mouth is acted on exclusively by the great zygomatic muscles, which serve to draw the corners backwards and upwards; but judging from the manner in which the upper teeth are always exposed during laughter and broad smiling, as well as from my own sensations, I cannot doubt that some of the muscles running to the upper lip are likewise brought into moderate action. The upper and lower orbicular muscles of the eyes are at the same time more or less contracted; and there is an intimate connection, as explained in the chapter on weeping, between the orbiculars, especially the lower ones, and some of the muscles running to the upper lip. Henle remarks [10] on this head, that when a man closely shuts one eye he cannot avoid retracting the upper lip on the same side; conversely, if any one will place his finger on his lower eyelid, and then uncover his upper incisors as much as possible, he will feel, as his upper lip is drawn strongly upwards, that the muscles of the lower eyelid contract. In Henle's drawing, given in woodcut, fig. 2, the *musculus malaris* (H) which runs to the upper lip may be seen to form an almost integral part of the lower orbicular muscle.

Dr. Duchenne has given a large photograph of an old man (reduced on Plate III. fig 4), in his usual passive condition, and another of the same man (fig. 5), naturally smiling. The latter was instantly recognized by every one to whom it was shown as true to nature. He has also given, as an example of an unnatural or false smile, another photograph (fig. 6) of the same old man, with the corners of his mouth strongly retracted by the galvanization of the great zygomatic muscles. That the expression is not natural is clear, for I showed this

---

[10] Handbuch der System. Anat. des Menschen, 1858, B. i. s. 144. See my woodcut (H. fig. 2).

photograph to twenty-four persons, of whom three could not in the least tell what was meant, whilst the others, though they perceived that the expression was of the nature of a smile, answered in such words as "a wicked joke," "trying to laugh," "grinning laughter," "half-amazed laughter," &c. Dr. Duchenne attributes the falseness of the expression altogether to the orbicular muscles of the lower eyelids not being sufficiently contracted; for he justly lays great stress on their contraction in the expression of joy. No doubt there is much truth in this view, but not, as it appears to me, the whole truth. The contraction of the lower orbiculars is always accompanied, as we have seen, by the drawing up of the upper lip. Had the upper lip, in fig. 6, been thus acted on to a slight extent, its curvature would have been less rigid, the naso-labial furrow would have been slightly different, and the whole expression would, as I believe, have been more natural, independently of the more conspicuous effect from the stronger contraction of the lower eyelids. The corrugator muscle, moreover, in fig. 6, is too much contracted, causing a frown; and this muscle never acts under the influence of joy except during strongly pronounced or violent laughter.

By the drawing backwards and upwards of the corners of the mouth, through the contraction of the great zygomatic muscles, and by the raising of the upper lip, the cheeks are drawn upwards. Wrinkles are thus formed under the eyes, and, with old people, at their outer ends; and these are highly characteristic of laughter or smiling. As a gentle smile increases into a strong one, or into a laugh, every one may feel and see, if he will attend to his own sensations and look at himself in a mirror, that as the upper lip is drawn up and the lower orbiculars contract, the wrinkles in the lower eyelids and those beneath the eyes are much strengthened or increased. At the same time, as I have repeatedly observed, the eyebrows are slightly lowered, which shows that the upper as well as the lower orbiculars contract at least to some degree, though this passes unperceived, as far as our sensations are concerned. If the original photograph of the old man, with his countenance in its usual placid state (fig. 4), be compared with that (fig. 5) in which he is naturally smiling, it may be seen that the eyebrows in the latter are a little lowered. I presume that this is owing to the upper orbiculars being impelled, through the force of long-associated habit, to act to a certain extent in concert with the lower orbiculars, which themselves contract in connection with the drawing up of the upper lip.

The tendency in the zygomatic muscles to contract under pleasurable emotions is shown by a curious fact, communicated to me by Dr. Browne, with respect to patients suffering from *general paralysis of the insane*.[11] "In this malady there is almost invariably optimism—delusions as to wealth, rank, grandeur—insane joyousness, benevolence, and profusion, while its very earliest physical symptom is trembling at the corners of the mouth and at the outer corners of the eyes. This is a well-recognized fact. Constant tremulous agitation of the inferior palpebral and great zygomatic muscles is pathognomic of the earlier stages of general paralysis. The countenance has a pleased and benevolent expression. As the disease advances other muscles become involved, but until complete fatuity is reached, the prevailing expression is that of feeble benevolence."

As in laughing and broadly smiling the cheeks and upper lip are much raised, the nose appears to be short-

[11] See, also, remarks to the same effect by Dr. J. Crichton Browne in 'Journal of Mental Science,' April, 1871, p. 149.

ened, and the skin on the bridge becomes finely wrinkled in transverse lines, with other oblique longitudinal lines on the sides. The upper front teeth are commonly exposed. A well-marked naso-labial fold is formed, which runs from the wing of each nostril to the corner of the mouth; and this fold is often double in old persons.

A bright and sparkling eye is as characteristic of a pleased or amused state of mind, as is the retraction of the corners of the mouth and upper lip with the wrinkles thus produced. Even the eyes of microcephalous idiots, who are so degraded that they never learn to speak, brighten slightly when they are pleased.[12] Under extreme laughter the eyes are too much suffused with tears to sparkle; but the moisture squeezed out of the glands during moderate laughter or smiling may aid in giving them lustre; though this must be of altogether subordinate importance, as they become dull from grief, though they are then often moist. Their brightness seems to be chiefly due to their tenseness,[13] owing to the contraction of the orbicular muscles and to the pressure of the raised cheeks. But, according to Dr. Piderit, who has discussed this point more fully than any other writer,[14] the tenseness may be largely attributed to the eyeballs becoming filled with blood and other fluids, from the acceleration of the circulation, consequent on the excitement of pleasure. He remarks on the contrast in the appearance of the eyes of a hectic patient with a rapid circulation, and of a man suffering from cholera with almost all the fluids of his body drained from him. Any cause which lowers the circulation deadens the eye. I remember seeing a man utterly

---

[12] C. Vogt, 'Mémoire sur les Microcéphales,' 1867, p. 21.
[13] Sir C. Bell, 'Anatomy of Expression,' p. 133.
[14] 'Mimik und Physiognomik,' 1867, s. 63--67.

prostrated by prolonged and severe exertion during a very hot day, and a bystander compared his eyes to those of a boiled codfish.

To return to the sounds produced during laughter. We can see in a vague manner how the utterance of sounds of some kind would naturally become associated with a pleasurable state of mind; for throughout a large part of the animal kingdom vocal or instrumental sounds are employed either as a call or as a charm by one sex for the other. They are also employed as the means for a joyful meeting between the parents and their offspring, and between the attached members of the same social community. But why the sounds which man utters when he is pleased have the peculiar reiterated character of laughter we do not know. Nevertheless we can see that they would naturally be as different as possible from the screams or cries of distress; and as in the production of the latter, the expirations are prolonged and continuous, with the inspirations short and interrupted, so it might perhaps have been expected with the sounds uttered from joy, that the expirations would have been short and broken with the inspirations prolonged; and this is the case.

It is an equally obscure point why the corners of the mouth are retracted and the upper lip raised during ordinary laughter. The mouth must not be opened to its utmost extent, for when this occurs during a paroxysm of excessive laughter hardly any sound is emitted; or it changes its tone and seems to come from deep down in the throat. The respiratory muscles, and even those of the limbs, are at the same time thrown into rapid vibratory movements. The lower jaw often partakes of this movement, and this would tend to prevent the mouth from being widely opened. But as a full volume of sound has to be poured forth, the orifice of the mouth must be large; and it is perhaps to gain this end that the corners are retracted and the upper lip raised. Although we can hardly account for the shape of the mouth during laughter, which leads to wrinkles being formed beneath the eyes, nor for the peculiar reiterated sound of laughter, nor for the quivering of the jaws, nevertheless we may infer that all these effects are due to some common cause. For they are all characteristic and expressive of a pleased state of mind in various kinds of monkeys.

A graduated series can be followed from violent to moderate laughter, to a broad smile, to a gentle smile, and to the expression of mere cheerfulness. During excessive laughter the whole body is often thrown backward and shakes, or is almost convulsed; the respiration is much disturbed; the head and face become gorged with blood, with the veins distended; and the orbicular muscles are spasmodically contracted in order to protect the eyes. Tears are freely shed. Hence, as formerly remarked, it is scarcely possible to point out any difference between the tear-stained face of a person after a paroxysm of excessive laughter and after a bitter crying-fit.[15] It is probably due to the close similarity of the spasmodic movements caused by these widely different emotions that hysteric patients alternately cry and laugh with violence, and that young children sometimes pass suddenly from the one to the other state. Mr. Swinhoe informs me that he has often seen the Chinese, when suffering from deep grief, burst out into hysterical fits of laughter.

---

[15] Sir J. Reynolds remarks ('Discourses,' xii. p. 100), "It is curious to observe, and it is certainly true, that the extremes of contrary passions are, with very little variation, expressed by the same action." He gives as an instance the frantic joy of a Bacchante and the grief of a Mary Magdalen.

I was anxious to know whether tears are freely shed during excessive laughter by most of the races of men, and I hear from my correspondents that this is the case. One instance was observed with the Hindoos, and they themselves said that it often occurred. So it is with the Chinese. The women of a wild tribe of Malays in the Malacca peninsula, sometimes shed tears when they laugh heartily, though this seldom occurs. With the Dyaks of Borneo it must frequently be the case, at least with the women, for I hear from the Rajah C. Brooke that it is a common expression with them to say "we nearly made tears from laughter." The aborigines of Australia express their emotions freely, and they are described by my correspondents as jumping about and clapping their hands for joy, and as often roaring with laughter. No less than four observers have seen their eyes freely watering on such occasions; and in one instance the tears rolled down their cheeks. Mr. Bulmer, a missionary in a remote part of Victoria, remarks, "that they have a keen sense of the ridiculous; they are excellent mimics, and when one of them is able to imitate the peculiarities of some absent member of the tribe, it is very common to hear all in the camp convulsed with laughter." With Europeans hardly anything excites laughter so easily as mimicry; and it is rather curious to find the same fact with the savages of Australia, who constitute one of the most distinct races in the world.

In Southern Africa with two tribes of Kafirs, especially with the women, their eyes often fill with tears during laughter. Gaika, the brother of the chief Sandilli, answers my query on this head, with the words, "Yes, that is their common practice." Sir Andrew Smith has seen the painted face of a Hottentot woman all furrowed with tears after a fit of laughter. In Northern Africa, with the Abyssinians, tears are secreted under the same circumstances. Lastly, in North America, the same fact has been observed in a remarkably savage and isolated tribe, but chiefly with the women; in another tribe it was observed only on a single occasion.

Excessive laughter, as before remarked, graduates into moderate laughter. In this latter case the muscles round the eyes are much less contracted, and there is little or no frowning. Between a gentle laugh and a broad smile there is hardly any difference, excepting that in smiling no reiterated sound is uttered, though a single rather strong expiration, or slight noise—a rudiment of a laugh—may often be heard at the commencement of a smile. On a moderately smiling countenance the contraction of the upper orbicular muscles can still just be traced by a slight lowering of the eyebrows. The contraction of the lower orbicular and palpebral muscles is much plainer, and is shown by the wrinkling of the lower eyelids and of the skin beneath them, together with a slight drawing up of the upper lip. From the broadest smile we pass by the finest steps into the gentlest one. In this latter case the features are moved in a much less degree, and much more slowly, and the mouth is kept closed. The curvature of the naso-labial furrow is also slightly different in the two cases. We thus see that no abrupt line of demarcation can be drawn between the movement of the features during the most violent laughter and a very faint smile.[16]

A smile, therefore, may be said to be the first stage in the development of a laugh. But a different and more probable view may be suggested; namely, that the habit of uttering loud reiterated sounds from a sense of pleasure, first led to the retraction of the corners of the mouth and of the upper lip, and to the contraction

---

[16] Dr. Piderit has come to the same conclusion, ibid. s. 99.

of the orbicular muscles; and that now, through association and long-continued habit, the same muscles are brought into slight play whenever any cause excites in us a feeling which, if stronger, would have led to laughter; and the result is a smile.

Whether we look at laughter as the full development of a smile, or, as is more probable, at a gentle smile as the last trace of a habit, firmly fixed during many generations, of laughing whenever we are joyful, we can follow in our infants the gradual passage of the one into the other. It is well known to those who have the charge of young infants, that it is difficult to feel sure when certain movements about their mouths are really expressive; that is, when they really smile. Hence I carefully watched my own infants. One of them at the age of forty-five days, and being at the time in a happy frame of mind, smiled; that is, the corners of the mouth were retracted, and simultaneously the eyes became decidedly bright. I observed the same thing on the following day; but on the third day the child was not quite well and there was no trace of a smile, and this renders it probable that the previous smiles were real. Eight days subsequently and during the next succeeding week, it was remarkable how his eyes brightened whenever he smiled, and his nose became at the same time transversely wrinkled. This was now accompanied by a little bleating noise, which perhaps represented a laugh. At the age of 113 days these little noises, which were always made during expiration, assumed a slightly different character, and were more broken or interrupted, as in sobbing; and this was certainly incipient laughter. The change in tone seemed to me at the time to be connected with the greater lateral extension of the mouth as the smiles became broader.

In a second infant the first real smile was observed

at about the same age, viz. forty-five days; and in a third, at a somewhat earlier age. The second infant, when sixty-five days old, smiled much more broadly and plainly than did the one first mentioned at the same age; and even at this early age uttered noises very like laughter. In this gradual acquirement, by infants, of the habit of laughing, we have a case in some degree analogous to that of weeping. As practice is requisite with the ordinary movements of the body, such as walking, so it seems to be with laughing and weeping. The art of screaming, on the other hand, from being of service to infants, has become finely developed from the earliest days.

High spirits, cheerfulness.—A man in high spirits, though he may not actually smile, commonly exhibits some tendency to the retraction of the corners of his mouth. From the excitement of pleasure, the circulation becomes more rapid; the eyes are bright, and the colour of the face rises. The brain, being stimulated by the increased flow of blood, reacts on the mental powers; lively ideas pass still more rapidly through the mind, and the affections are warmed. I heard a child, a little under four years old, when asked what was meant by being in good spirits, answer, "It is laughing, talking, and kissing." It would be difficult to give a truer and more practical definition. A man in this state holds his body erect, his head upright, and his eyes open. There is no drooping of the features, and no contraction of the eyebrows. On the contrary, the frontal muscle, as Moreau observes,[17] tends to contract slightly; and this smooths the brow, removes every trace of a frown, arches

[17] 'La Physionomie,' par G. Lavater, edit. of 1820, vol. iv. p. 224. See, also, Sir C. Bell, 'Anatomy of Expression,' p. 172, for the quotation given below.

the eyebrows a little, and raises the eyelids. Hence the Latin phrase, *exporrigere frontem*—to unwrinkle the brow—means, to be cheerful or merry. The whole expression of a man in good spirits is exactly the opposite of that of one suffering from sorrow. According to Sir C. Bell, "In all the exhilarating emotions the eyebrows, eyelids, the nostrils, and the angles of the mouth are raised. In the depressing passions it is the reverse." Under the influence of the latter the brow is heavy, the eyelids, cheeks, mouth, and whole head droop; the eyes are dull; the countenance pallid, and the respiration slow. In joy the face expands, in grief it lengthens. Whether the principle of antithesis has here come into play in producing these opposite expressions, in aid of the direct causes which have been specified and which are sufficiently plain, I will not pretend to say.

With all the races of man the expression of good spirit appears to be the same, and is easily recognized. My informants, from various parts of the Old and New Worlds, answer in the affirmative to my queries on this head, and they give some particulars with respect to Hindoos, Malays, and New Zealanders. The brightness of the eyes of the Australians has struck four observers, and the same fact has been noticed with Hindoos, New Zealanders, and the Dyaks of Borneo.

Savages sometimes express their satisfaction not only by smiling, but by gestures derived from the pleasure of eating. Thus Mr. Wedgwood [18] quotes Petherick that the negroes on the Upper Nile began a general rubbing of their bellies when he displayed his beads; and Leichhardt says that the Australians smacked and clacked their mouths at the sight of his horses and bullocks, and

---

[18] A 'Dictionary of English Etymology,' 2nd edit. 1872, Introduction, p. xliv.

more especially of his kangaroo dogs. The Greenlanders, "when they affirm anything with pleasure, suck down air with a certain sound;"[19] and this may be an imitation of the act of swallowing savoury food.

Laughter is suppressed by the firm contraction of the orbicular muscles of the mouth, which prevents the great zygomatic and other muscles from drawing the lips backwards and upwards. The lower lip is also sometimes held by the teeth, and this gives a roguish expression to the face, as was observed with the blind and deaf Laura Bridgman.[20] The great zygomatic muscle is sometimes variable in its course, and I have seen a young woman in whom the *depressores anguli oris* were brought into strong action in suppressing a smile; but this by no means gave to her countenance a melancholy expression, owing to the brightness of her eyes.

Laughter is frequently employed in a forced manner to conceal or mask some other state of mind, even anger. We often see persons laughing in order to conceal their shame or shyness. When a person purses up his mouth, as if to prevent the possibility of a smile, though there is nothing to excite one, or nothing to prevent its free indulgence, an affected, solemn, or pedantic expression is given; but of such hybrid expressions nothing more need here be said. In the case of derision, a real or pretended smile or laugh is often blended with the expression proper to contempt, and this may pass into angry contempt or scorn. In such cases the meaning of the laugh or smile is to show the offending person that he excites only amusement.

Love, tender feelings, &c.—Although the emotion of

---

[19] Crantz, quoted by Tylor, 'Primitive Culture,' 1871, vol. i. p. 169.

[20] F. Lieber, 'Smithsonian Contributions,' 1851, vol. ii. p. 7.

love, for instance that of a mother for her infant, is one of the strongest of which the mind is capable, it can hardly be said to have any proper or peculiar means of expression; and this is intelligible, as it has not habitually led to any special line of action. No doubt, as affection is a pleasurable sensation, it generally causes a gentle smile and some brightening of the eyes. A strong desire to touch the beloved person is commonly felt; and love is expressed by this means more plainly than by any other.[21] Hence we long to clasp in our arms those whom we tenderly love. We probably owe this desire to inherited habit, in association with the nursing and tending of our children, and with the mutual caresses of lovers.

With the lower animals we see the same principle of pleasure derived from contact in association with love. Dogs and cats manifestly take pleasure in rubbing against their masters and mistresses, and in being rubbed or patted by them. Many kinds of monkeys, as I am assured by the keepers in the Zoological Gardens, delight in fondling and being fondled by each other, and by persons to whom they are attached. Mr. Bartlett has described to me the behaviour of two chimpanzees, rather older animals than those generally imported into this country, when they were first brought together. They sat opposite, touching each other with their much protruded lips; and the one put his hand on the shoulder of the other. They then mutually folded each other in their arms. Afterwards they stood up, each with one arm on the shoulder of the other, lifted up their heads, opened their mouths, and yelled with delight.

---

[21] Mr. Bain remarks ('Mental and Moral Science,' 1868, p. 239), "Tenderness is a pleasurable emotion, variously stimulated, whose effort is to draw human beings into mutual embrace."

We Europeans are so accustomed to kissing as a mark of affection, that it might be thought to be innate in mankind; but this is not the case. Steele was mistaken when he said "Nature was its author, and it began with the first courtship." Jemmy Button, the Fuegian, told me that this practice was unknown in his land. It is equally unknown with the New Zealanders, Tahitians, Papuans, Australians, Somals of Africa, and the Esquimaux.[22] But it is so far innate or natural that it apparently depends on pleasure from close contact with a beloved person; and it is replaced in various parts of the world, by the rubbing of noses, as with the New Zealanders and Laplanders, by the rubbing or patting of the arms, breasts, or stomachs, or by one man striking his own face with the hands or feet of another. Perhaps the practice of blowing, as a mark of affection, on various parts of the body may depend on the same principle.[23]

The feelings which are called tender are difficult to analyse; they seem to be compounded of affection, joy, and especially of sympathy. These feelings are in themselves of a pleasurable nature, excepting when pity is too deep, or horror is aroused, as in hearing of a tortured man or animal. They are remarkable under our present point of view from so readily exciting the secretion of tears. Many a father and son have wept on meeting after a long separation, especially if the meeting has been unexpected. No doubt extreme joy by itself tends to act on the lacrymal glands; but on such occasions as the foregoing vague thoughts of the grief which would have

[22] Sir J. Lubbock, 'Prehistoric Times,' 2nd edit. 1869, p. 552, gives full authorities for these statements. The quotation from Steele is taken from this work.

[23] See a full acount, with references, by E. B. Tylor, 'Researches into the Early History of Mankind,' 2nd edit. 1870, p. 51.

been felt had the father and son never met, will probably have passed through their minds; and grief naturally leads to the secretion of tears. Thus on the return of Ulysses:—

> "Telemachus
> Rose, and clung weeping round his father's breast.
> There the pent grief rained o'er them, yearning thus.
> * * * * * *
> Thus piteously they wailed in sore unrest,
> And on their weepings had gone down the day,
> But that at last Telemachus found words to say."
> *Worsley's Translation of the Odyssey,*
> Book xvi. st. 27.

So again when Penelope at last recognized her husband:—

> "Then from her eyelids the quick tears did start
> And she ran to him from her place, and threw
> Her arms about his neck, and a warm dew
> Of kisses poured upon him, and thus spake: "
> Book xxiii. st. 27.

The vivid recollection of our former home, or of long-past happy days, readily causes the eyes to be suffused with tears; but here, again, the thought naturally occurs that these days will never return. In such cases we may be said to sympathize with ourselves in our present, in comparison with our former, state. Sympathy with the distresses of others, even with the imaginary distresses of a heroine in a pathetic story, for whom we feel no affection, readily excites tears. So does sympathy with the happiness of others, as with that of a lover, at last successful after many hard trials in a well-told tale.

Sympathy appears to constitute a separate or distinct emotion; and it is especially apt to excite the lacrymal glands. This holds good whether we give or receive sympathy. Every one must have noticed how readily children burst out crying if we pity them for some small hurt. With the melancholic insane, as Dr. Crichton Browne informs me, a kind word will often plunge them into unrestrained weeping. As soon as we express our pity for the grief of a friend, tears often come into our own eyes. The feeling of sympathy is commonly explained by assuming that, when we see or hear of suffering in another, the idea of suffering is called up so vividly in our own minds that we ourselves suffer. But this explanation is hardly sufficient, for it does not account for the intimate alliance between sympathy and affection. We undoubtedly sympathize far more deeply with a beloved than with an indifferent person; and the sympathy of the one gives us far more relief than that of the other. Yet assuredly we can sympathize with those for whom we feel no affection.

Why suffering, when actually experienced by ourselves, excites weeping, has been discussed in a former chapter. With respect to joy, its natural and universal expression is laughter; and with all the races of man loud laughter leads to the secretion of tears more freely than does any other cause excepting distress. The suffusion of the eyes with tears, which undoubtedly occurs under great joy, though there is no laughter, can, as it seems to me, be explained through habit and association on the same principles as the effusion of tears from grief, although there is no screaming. Nevertheless it is not a little remarkable that sympathy with the distresses of others should excite tears more freely than our own distress; and this certainly is the case. Many a man, from whose eyes no suffering of his own could wring a tear, has shed tears at the sufferings of a beloved friend. It is still more remarkable that sympathy with the happiness or good fortune of those whom we tenderly love should lead to the same result, whilst a similar happiness felt by ourselves would leave our eyes dry. We should, however, bear in mind that the long-continued habit of restraint which is so powerful in checking the free flow of tears from bodily pain, has not been brought into play in preventing a moderate effusion of tears in sympathy with the sufferings or happiness of others.

Music has a wonderful power, as I have elsewhere attempted to show,[24] of recalling in a vague and indefinite manner, those strong emotions which were felt during long-past ages, when, as is probable, our early progenitors courted each other by the aid of vocal tones. And as several of our strongest emotions—grief, great joy, love, and sympathy—lead to the free secretion of tears, it is not surprising that music should be apt to cause our eyes to become suffused with tears, especially when we are already softened by any of the tenderer feelings. Music often produces another peculiar effect. We know that every strong sensation, emotion, or excitement—extreme pain, rage, terror, joy, or the passion of love—all have a special tendency to cause the muscles to tremble; and the thrill or slight shiver which runs down the backbone and limbs of many persons when they are powerfully affected by music, seems to bear the same relation to the above trembling of the body, as a slight suffusion of tears from the power of music does to weeping from any strong and real emotion.

Devotion.—As devotion is, in some degree, related to affection, though mainly consisting of reverence, often combined with fear, the expression of this state of mind may here be briefly noticed. With some sects, both past and present, religion and love have been strangely combined; and it has even been maintained, lamentable

[24] 'The Descent of Man,' vol. ii. p. 336.

as the fact may be, that the holy kiss of love differs but little from that which a man bestows on a woman, or a woman on a man.[25] Devotion is chiefly expressed by the face being directed towards the heavens, with the eyeballs upturned. Sir C. Bell remarks that, at the approach of sleep, or of a fainting-fit, or of death, the pupils are drawn upwards and inwards; and he believes that "when we are wrapt in devotional feelings, and outward impressions are unheeded, the eyes are raised by an action neither taught nor acquired;" and that this is due to the same cause as in the above cases.[26] That the eyes are upturned during sleep is, as I hear from Professor Donders, certain. With babies, whilst sucking their mother's breast, this movement of the eyeballs often gives to them an absurd appearance of ecstatic delight; and here it may be clearly perceived that a struggle is going on against the position naturally assumed during sleep. But Sir C. Bell's explanation of the fact, which rests on the assumption that certain muscles are more under the control of the will than others is, as I hear from Professor Donders, incorrect. As the eyes are often turned up in prayer, without the mind being so much absorbed in thought as to approach to the unconsciousness of sleep, the movement is probably a conventional one—the result of the common belief that Heaven, the source of Divine power to which we pray, is seated above us.

A humble kneeling posture, with the hands upturned and palms joined, appears to us, from long habit, a gesture so appropriate to devotion, that it might be thought to be innate; but I have not met with any evidence to

---

[25] Dr. Maudsley has a discussion to this effect in his 'Body and Mind,' 1870, p. 85.

[26] 'The Anatomy of Expression,' p. 103, and 'Philosophical Transactions,' 1823, p. 182.

this effect with the various extra-European races of mankind. During the classical period of Roman history it does not appear, as I hear from an excellent classic, that the hands were thus joined during prayer. Mr. Hensleigh Wedgwood has apparently given [27] the true explanation, though this implies that the attitude is one of slavish subjection. "When the suppliant kneels and holds up his hands with the palms joined, he represents a captive who proves the completeness of his submission by offering up his hands to be bound by the victor. It is the pictorial representation of the Latin *dare manus*, to signify submission." Hence it is not probable that either the uplifting of the eyes or the joining of the open hands, under the influence of devotional feelings, are innate or truly expressive actions; and this could hardly have been expected, for it is very doubtful whether feelings, such as we should now rank as devotional, affected the hearts of men, whilst they remained during past ages in an uncivilized condition.

---

[27] 'The Origin of Language,' 1866, p. 146. Mr. Tylor ('Early History of Mankind,' 2nd edit. 1870, p. 48) gives a more complex origin to the position of the hands during prayer.

CHAPTER IX.

REFLECTION—MEDITATION—ILL-TEMPER—SULKINESS —DETERMINATION.

The act of frowning—Reflection with an effort, or with the perception of something difficult or disagreeable—Abstracted meditation—Ill-temper—Moroseness—Obstinacy—Sulkiness and pouting—Decision or determination—The firm closure of the mouth.

THE corrugators, by their contraction, lower the eyebrows and bring them together, producing vertical furrows on the forehead—that is, a frown. Sir C. Bell, who erroneously thought that the corrugator was peculiar to man, ranks it as "the most remarkable muscle of the human face. It knits the eyebrows with an energetic effort, which unaccountably, but irresistibly, conveys the idea of mind." Or, as he elsewhere says, "when the eyebrows are knit, energy of mind is apparent, and there is the mingling of thought and emotion with the savage and brutal rage of the mere animal."[1] There

[1] 'Anatomy of Expression,' pp. 137, 139. It is not surprising that the corrugators should have become much more developed in man than in the anthropoid apes; for they are brought into incessant action by him under various circumstances, and will have been strengthened and modified by the inherited effects of use. We have seen how important a part they play, together with the orbiculares, in protecting the eyes from being too much gorged with blood during violent expiratory movements. When the eyes are closed as quickly and as forcibly as possible,

is much truth in these remarks, but hardly the whole truth. Dr. Duchenne has called the corrugator the muscle of reflection;[2] but this name, without some limitation, cannot be considered as quite correct.

A man may be absorbed in the deepest thought, and his brow will remain smooth until he encounters some obstacle in his train of reasoning, or is interrupted by some disturbance, and then a frown passes like a shadow over his brow. A half-starved man may think intently how to obtain food, but he probably will not frown unless he encounters either in thought or action some difficulty, or finds the food when obtained nauseous. I have noticed that almost everyone instantly frowns if he perceives a strange or bad taste in what he is eating. I asked several persons, without explaining my object, to listen intently to a very gentle tapping sound, the nature and source of which they all perfectly knew, and not one frowned; but a man who joined us, and who could not conceive what we were all doing in profound silence, when asked to listen, frowned much, though not in an ill-temper, and said he could not in the least understand what we all wanted. Dr. Piderit,[3] who has published remarks to the same effect, adds that stammerers generally frown in speaking; and that a man in doing even so trifling a thing as pulling on a boot, frowns if

to save them from being injured by a blow, the corrugators contract. With savages or other men whose heads are uncovered, the eyebrows are continually lowered and contracted to serve as a shade against a too strong light; and this is effected partly by the corrugators. This movement would have been more especially serviceable to man, as soon as his early progenitors held their heads erect. Lastly, Prof. Donders believes ('Archives of Medicine,' ed. by L. Beale, 1870, vol. v. p. 34), that the corrugators are brought into action in causing the eyeball to advance in accommodation for proximity in vision.

[2] 'Mécanisme de la Physionomie Humaine,' Album, Légende iii.

[3] 'Mimik und Physiognomik,' s. 46.

he finds it too tight. Some persons are such habitual frowners, that the mere effort of speaking almost always causes their brows to contract.

Men of all races frown when they are in any way perplexed in thought, as I infer from the answers which I have received to my queries; but I framed them badly, confounding absorbed meditation with perplexed reflection. Nevertheless, it is clear that the Australians, Malays, Hindoos, and Kafirs of South Africa frown, when they are puzzled. Dobritzhoffer remarks that the Guaranies of South America on like occasions knit their brows.[4]

From these considerations, we may conclude that frowning is not the expression of simple reflection, however profound, or of attention, however close, but of something difficult or displeasing encountered in a train of thought or in action. Deep reflection can, however, seldom be long carried on without some difficulty, so that it will generally be accompanied by a frown. Hence it is that frowning commonly gives to the countenance, as Sir C. Bell remarks, an aspect of intellectual energy. But in order that this effect may be produced, the eyes must be clear and steady, or they may be cast downwards, as often occurs in deep thought. The countenance must not be otherwise disturbed, as in the case of an ill-tempered or peevish man, or of one who shows the effects of prolonged suffering, with dulled eyes and drooping jaw, or who perceives a bad taste in his food, or who finds it difficult to perform some trifling act, such as threading a needle. In these cases a frown may often be seen, but it will be accompanied by some other expression, which will entirely prevent the countenance hav-

---

[4] 'History of the Abipones,' Eng. translat. vol. ii. p. 59, as quoted by Lubbock, 'Origin of Civilisation,' 1870, p. 355.

ing an appearance of intellectual energy or of profound thought.

We may now inquire how it is that a frown should express the perception of something difficult or disagreeable, either in thought or action. In the same way as naturalists find it advisable to trace the embryological development of an organ in order fully to understand its structure, so with the movements of expression it is advisable to follow as nearly as possible the same plan. The earliest and almost sole expression seen during the first days of infancy, and then often exhibited, is that displayed during the act of screaming; and screaming is excited, both at first and for some time afterwards, by every distressing or displeasing sensation and emotion,—by hunger, pain, anger, jealousy, fear, &c. At such times the muscles round the eyes are strongly contracted; and this, as I believe, explains to a large extent the act of frowning during the remainder of our lives. I repeatedly observed my own infants, from under the age of one week to that of two or three months, and found that when a screaming-fit came on gradually, the first sign was the contraction of the corrugators, which produced a slight frown, quickly followed by the contraction of the other muscles round the eyes. When an infant is uncomfortable or unwell, little frowns—as I record in my notes—may be seen incessantly passing like shadows over its face; these being generally, but not always, followed sooner or later by a crying-fit. For instance, I watched for some time a baby, between seven and eight weeks old, sucking some milk which was cold, and therefore displeasing to him; and a steady little frown was maintained all the time. This was never developed into an actual crying-fit, though occasionally every stage of close approach could be observed.

As the habit of contracting the brows has been followed by infants during innumerable generations, at the commencement of every crying or screaming fit, it has become firmly associated with the incipient sense of something distressing or disagreeable. Hence under similar circumstances it would be apt to be continued during maturity, although never then developed into a crying-fit. Screaming or weeping begins to be voluntarily restrained at an early period of life, whereas frowning is hardly ever restrained at any age. It is perhaps worth notice that with children much given to weeping, anything which perplexes their minds, and which would cause most other children merely to frown, readily makes them weep. So with certain classes of the insane, any effort of mind, however slight, which with an habitual frowner would cause a slight frown, leads to their weeping in an unrestrained manner. It is not more surprising that the habit of contracting the brows at the first perception of something distressing, although gained during infancy, should be retained during the rest of our lives, than that many other associated habits acquired at an early age should be permanently retained both by man and the lower animals. For instance, full-grown cats, when feeling warm and comfortable, often retain the habit of alternately protruding their fore-feet with extended toes, which habit they practised for a definite purpose whilst sucking their mothers.

Another and distinct cause has probably strengthened the habit of frowning, whenever the mind is intent on any subject and encounters some difficulty. Vision is the most important of all the senses, and during primeval times the closest attention must have been incessantly directed towards distant objects for the sake of obtaining prey and avoiding danger. I remember being struck, whilst travelling in parts of South America, which were dangerous from the presence of Indians, how incessantly, yet as it appeared unconsciously, the half-wild Gauchos closely scanned the whole horizon. Now, when any one with no covering on his head (as must have been aboriginally the case with mankind), strives to the utmost to distinguish in broad daylight, and especially if the sky is bright, a distant object, he almost invariably contracts his brows to prevent the entrance of too much light; the lower eyelids, cheeks, and upper lip being at the same time raised, so as to lessen the orifice of the eyes. I have purposely asked several persons, young and old, to look, under the above circumstances, at distant objects, making them believe that I only wished to test the power of their vision; and they all behaved in the manner just described. Some of them, also, put their open, flat hands over their eyes to keep out the excess of light. Gratiolet, after making some remarks to nearly the same effect,[5] says, "Ce sont là des attitudes de vision difficile." He concludes that the muscles round the eyes contract partly for the sake of excluding too much light (which appears to me the more important end), and partly to prevent all rays striking the retina, except those which come direct from the object that is scrutinized. Mr. Bowman, whom I consulted on this point, thinks that the contraction of the surrounding muscles may, in addition, "partly sustain the consensual movements of the two eyes, by giving a firmer support while the globes are brought to binocular vision by their own proper muscles."

As the effort of viewing with care under a bright light a distant object is both difficult and irksome, and

---

[5] 'De la Physionomie,' pp. 15, 144, 146. Mr. Herbert Spencer accounts for frowning exclusively by the habit of contracting the brows as a shade to the eyes in a bright light: see 'Principles of Physiology,' 2nd edit. 1872, p. 546.

as this effort has been habitually accompanied, during numberless generations, by the contraction of the eyebrows, the habit of frowning will thus have been much strengthened; although it was originally practised during infancy from a quite independent cause, namely as the first step in the protection of the eyes during screaming. There is, indeed, much analogy, as far as the state of the mind is concerned, between intently scrutinizing a distant object, and following out an obscure train of thought, or performing some little and troublesome mechanical work. The belief that the habit of contracting the brows is continued when there is no need whatever to exclude too much light, receives support from the cases formerly alluded to, in which the eyebrows or eyelids are acted on under certain circumstances in a useless manner, from having been similarly used, under analogous circumstances, for a serviceable purpose. For instance, we voluntarily close our eyes when we do not wish to see any object, and we are apt to close them, when we reject a proposition, as if we could not or would not see it; or when we think about something horrible. We raise our eyebrows when we wish to see quickly all round us, and we often do the same, when we earnestly desire to remember something; acting as if we endeavoured to see it.

Abstraction. Meditation.—When a person is lost in thought with his mind absent, or, as it is sometimes said, "when he is in a brown study," he does not frown, but his eyes appear vacant. The lower eyelids are generally raised and wrinkled, in the same manner as when a short-sighted person tries to distinguish a distant object; and the upper orbicular muscles are at the same time slightly contracted. The wrinkling of the lower eyelids under these circumstances has been observed with some savages, as by Mr. Dyson Lacy with the Australians of Queensland, and several times by Mr. Geach with the Malays of the interior of Malacca. What the meaning or cause of this action may be, cannot at present be explained; but here we have another instance of movement round the eyes in relation to the state of the mind.

The vacant expression of the eyes is very peculiar, and at once shows when a man is completely lost in thought. Professor Donders has, with his usual kindness, investigated this subject for me. He has observed others in this condition, and has been himself observed by Professor Engelmann. The eyes are not then fixed on any object, and therefore not, as I had imagined, on some distant object. The lines of vision of the two eyes even often become slightly divergent; the divergence, if the head be held vertically, with the plane of vision horizontal, amounting to an angle of 2° as a maximum. This was ascertained by observing the crossed double image of a distant object. When the head droops forward, as often occurs with a man absorbed in thought, owing to the general relaxation of his muscles, if the plane of vision be still horizontal, the eyes are necessarily a little turned upwards, and then the divergence is as much as 3°, or 3° 5′: if the eyes are turned still more upwards, it amounts to between 6° and 7°. Professor Donders attributes this divergence to the almost complete relaxation of certain muscles of the eyes, which would be apt to follow from the mind being wholly absorbed.[6] The active condition of the muscles of the eyes

[6] Gratiolet remarks (De la Phys. p. 35), "Quand l'attention est fixée sur quelque image intérieure, l'œil regarde dans le vide et s'associe automatiquement à la contemplation de l'esprit." But this view hardly deserves to be called an explanation.

is that of convergence; and Professor Donders remarks, as bearing on their divergence during a period of complete abstraction, that when one eye becomes blind, it almost always, after a short lapse of time, deviates outwards; for its muscles are no longer used in moving the eyeball inwards for the sake of binocular vision.

Perplexed reflection is often accompanied by certain movements or gestures. At such times we commonly raise our hands to our foreheads, mouths, or chins; but we do not act thus, as far as I have seen, when we are quite lost in meditation, and no difficulty is encountered. Plautus, describing in one of his plays [7] a puzzled man, says, "Now look, he has pillared his chin upon his hand." Even so trifling and apparently unmeaning a gesture as the raising of the hand to the face has been observed with some savages. M. J. Mansel Weale has seen it with the Kafirs of South Africa; and the native chief Gaika adds, that men then "sometimes pull their beards." Mr. Washington Matthews, who attended to some of the wildest tribes of Indians in the western regions of the United States, remarks that he has seen them when concentrating their thoughts, bring their "hands, usually the thumb and index finger, in contact with some part of the face, commonly the upper lip." We can understand why the forehead should be pressed or rubbed, as deep thought tries the brain; but why the hand should be raised to the mouth or face is far from clear.

Ill-temper.—We have seen that frowning is the natural expression of some difficulty encountered, or of something disagreeable experienced either in thought or action, and he whose mind is often and readily affected

[7] 'Miles Gloriosus,' act ii. sc. 2.

in this way, will be apt to be ill-tempered, or slightly angry, or peevish, and will commonly show it by frowning. But a cross expression, due to a frown, may be counteracted, if the mouth appears sweet, from being habitually drawn into a smile, and the eyes are bright and cheerful. So it will be if the eye is clear and steady, and there is the appearance of earnest reflection. Frowning, with some depression of the corners of the mouth, which is a sign of grief, gives an air of peevishness. If a child (see Plate IV., fig. 2) [8] frowns much whilst crying, but does not strongly contract in the usual manner the orbicular muscles, a well-marked expression of anger or even of rage, together with misery, is displayed.

If the whole frowning brow be drawn much downward by the contraction of the pyramidal muscles of the nose, which produces transverse wrinkles or folds across the base of the nose, the expression becomes one of moroseness. Duchenne believes that the contraction of this muscle, without any frowning, gives the appearance of extreme and aggressive hardness.[9] But I much doubt whether this is a true or natural expression. I have shown Duchenne's photograph of a young man, with this muscle strongly contracted by means of galvanism, to eleven persons, including some artists, and none of them could form an idea what was intended, except one, a girl, who answered correctly, "surely reserve." When I first looked at this photograph, knowing what was intended, my imagination added, as I believe, what was necessary, namely, a frowning brow; and consequently

---

[8] The original photograph by Herr Kindermann is much more expressive than this copy, as it shows the frown on the brow more plainly.

[9] 'Mécanisme de la Physionomie Humaine,' Album, Légende iv. figs. 16--18.

the expression appeared to me true and extremely morose.

A firmly closed mouth, in addition to a lowered and frowning brow, gives determination to the expression, or may make it obstinate and sullen. How it comes that the firm closure of the mouth gives the appearance of determination will presently be discussed. An expression of sullen obstinacy has been clearly recognized by my informants, in the natives of six different regions of Australia. It is well marked, according to Mr. Scott, with the Hindoos. It has been recognized with the Malays, Chinese, Kafirs, Abyssinians, and in a conspicuous degree, according to Dr. Rothrock, with the wild Indians of North America, and according to Mr. D. Forbes, with the Aymaras of Bolivia. I have also observed it with the Araucanos of southern Chili. Mr. Dyson Lacy remarks that the natives of Australia, when in this frame of mind, sometimes fold their arms across their breasts, an attitude which may be seen with us. A firm determination, amounting to obstinacy, is, also, sometimes expressed by both shoulders being kept raised, the meaning of which gesture will be explained in the following chapter.

With young children sulkiness is shown by pouting, or, as it is sometimes called, "making a snout." [10] When the corners of the mouth are much depressed, the lower lip is a little everted and protruded; and this is likewise called a pout. But the pouting here referred to, consists of the protrusion of both lips into a tubular form, sometimes to such an extent as to project as far as the end of the nose, if this be short. Pouting is generally accompanied by frowning, and sometimes by the

---

[10] Hensleigh Wedgwood on 'The Origin of Language,' 1866, p. 78.

utterance of a booing or whooing noise. This expression is remarkable, as almost the sole one, as far as I know, which is exhibited much more plainly during childhood, at least with Europeans, than during maturity. There is, however, some tendency to the protrusion of the lips with the adults of all races under the influence of great rage. Some children pout when they are shy, and they can then hardly be called sulky.

From inquiries which I have made in several large families, pouting does not seem very common with European children; but it prevails throughout the world, and must be both common and strongly marked with most savage races, as it has caught the attention of many observers. It has been noticed in eight different districts of Australia; and one of my informants remarks how greatly the lips of the children are then protruded. Two observers have seen pouting with the children of Hindoos; three, with those of the Kafirs and Fingoes of South Africa, and with the Hottentots; and two, with the children of the wild Indians of North America. Pouting has also been observed with the Chinese, Abyssinians, Malays of Malacca, Dyaks of Borneo, and often with the New Zealanders. Mr. Mansel Weale informs me that he has seen the lips much protruded, not only with the children of the Kafirs, but with the adults of both sexes when sulky; and Mr. Stack has sometimes observed the same thing with the men, and very frequently with the women of New Zealand. A trace of the same expression may occasionally be detected even with adult Europeans.

We thus see that the protrusion of the lips, especially with young children, is characteristic of sulkiness throughout the greater part of the world. This movement apparently results from the retention, chiefly during youth, of a primordial habit, or from an occasional reversion to it. Young orangs and chimpanzees protrude their lips to an extraordinary degree, as described in a former chapter, when they are discontented, somewhat angry, or sulky; also when they are surprised, a little frightened, and even when slightly pleased. Their mouths are protruded apparently for the sake of making the various noises proper to these several states of mind; and its shape, as I observed with the chimpanzee, differed slightly when the cry of pleasure and that of anger were uttered. As soon as these animals become enraged, the shape of the mouth wholly changes, and the teeth are exposed. The adult orang when wounded is said to emit "a singular cry, consisting at first of high notes, which at length deepen into a low roar. While giving out the high notes he thrusts out his lips into a funnel shape, but in uttering the low notes he holds his mouth wide open."[11] With the gorilla, the lower lip is said to be capable of great elongation. If then our semi-human progenitors protruded their lips when sulky or a little angered, in the same manner as do the existing anthropoid apes, it is not an anomalous, though a curious fact, that our children should exhibit, when similarly affected, a trace of the same expression, together with some tendency to utter a noise. For it is not at all unusual for animals to retain, more or less perfectly, during early youth, and subsequently to lose, characters which were aboriginally possessed by their adult progenitors, and which are still retained by distinct species, their near relations.

Nor is it an anomalous fact that the children of savages should exhibit a stronger tendency to protrude their lips, when sulky, than the children of civilized

[11] Müller, as quoted by Huxley, 'Man's Place in Nature,' 1863, p. 38.

Europeans; for the essence of savagery seems to consist in the retention of a primordial condition, and this occasionally holds good even with bodily peculiarities.[12] It may be objected to this view of the origin of pouting, that the anthropoid apes likewise protrude their lips when astonished and even when a little pleased; whilst with us this expression is generally confined to a sulky frame of mind. But we shall see in a future chapter that with men of various races surprise does sometimes lead to a slight protrusion of the lips, though great surprise or astonishment is more commonly shown by the mouth being widely opened. As when we smile or laugh we draw back the corners of the mouth, we have lost any tendency to protrude the lips, when pleased, if indeed our early progenitors thus expressed pleasure.

A little gesture made by sulky children may here be noticed, namely, their "showing a cold shoulder." This has a different meaning, as, I believe, from the keeping both shoulders raised. A cross child, sitting on its parent's knee, will lift up the near shoulder, then jerk it away, as if from a caress, and afterwards give a backward push with it, as if to push away the offender. I have seen a child, standing at some distance from any one, clearly express its feelings by raising one shoulder, giving it a little backward movement, and then turning away its whole body.

Decision or determination.—The firm closure of the mouth tends to give an expression of determination or decision to the countenance. No determined man probably ever had an habitually gaping mouth. Hence, also, a small and weak lower jaw, which seems to indicate that

---

[12] I have given several instances in my 'Descent of Man,' vol. i. chap. iv.

the mouth is not habitually and firmly closed, is commonly thought to be characteristic of feebleness of character. A prolonged effort of any kind, whether of body or mind, implies previous determination; and if it can be shown that the mouth is generally closed with firmness before and during a great and continued exertion of the muscular system, then, through the principle of association, the mouth would almost certainly be closed as soon as any determined resolution was taken. Now several observers have noticed that a man, in commencing any violent muscular effort, invariably first distends his lungs with air, and then compresses it by the strong contraction of the muscles of the chest; and to effect this the mouth must be firmly closed. Moreover, as soon as the man is compelled to draw breath, he still keeps his chest as much distended as possible.

Various causes have been assigned for this manner of acting. Sir C. Bell maintains [13] that the chest is distended with air, and is kept distended at such times, in order to give a fixed support to the muscles which are thereto attached. Hence, as he remarks, when two men are engaged in a deadly contest, a terrible silence prevails, broken only by hard stifled breathing. There is silence, because to expel the air in the utterance of any sound would be to relax the support for the muscles of the arms. If an outcry is heard, supposing the struggle to take place in the dark, we at once know that one of the two has given up in despair.

Gratiolet admits [14] that when a man has to struggle with another to his utmost, or has to support a great weight, or to keep for a long time the same forced attitude, it is necessary for him first to make a deep inspira-

---

[13] 'Anatomy of Expression,' p. 190.
[14] 'De la Physionomie,' pp. 118--121.

tion, and then to cease breathing; but he thinks that Sir C. Bell's explanation is erroneous. He maintains that arrested respiration retards the circulation of the blood, of which I believe there is no doubt, and he adduces some curious evidence from the structure of the lower animals, showing, on the one hand, that a retarded circulation is necessary for prolonged muscular exertion, and, on the other hand, that a rapid circulation is necessary for rapid movements. According to this view, when we commence any great exertion, we close our mouths and stop breathing, in order to retard the circulation of the blood. Gratiolet sums up the subject by saying, "C'est là la vraie théorie de l'effort continu;" but how far this theory is admitted by other physiologists I do not know.

Dr. Piderit accounts[15] for the firm closure of the mouth during strong muscular exertion, on the principle that the influence of the will spreads to other muscles besides those necessarily brought into action in making any particular exertion; and it is natural that the muscles of respiration and of the mouth, from being so habitually used, should be especially liable to be thus acted on. It appears to me that there probably is some truth in this view, for we are apt to press the teeth hard together during violent exertion, and this is not requisite to prevent expiration, whilst the muscles of the chest are strongly contracted.

Lastly, when a man has to perform some delicate and difficult operation, not requiring the exertion of any strength, he nevertheless generally closes his mouth and ceases for a time to breathe; but he acts thus in order that the movements of his chest may not disturb those of his arms. A person, for instance, whilst threading a

---

[15] 'Mimik und Physiognomik,' s. 79.

needle, may be seen to compress his lips and either to stop breathing, or to breathe as quietly as possible. So it was, as formerly stated, with a young and sick chimpanzee, whilst it amused itself by killing flies with its knuckles, as they buzzed about on the window-panes. To perform an action, however trifling, if difficult, implies some amount of previous determination.

There appears nothing improbable in all the above assigned causes having come into play in different degrees, either conjointly or separately, on various occasions. The result would be a well-established habit, now perhaps inherited, of firmly closing the mouth at the commencement of and during any violent and prolonged exertion, or any delicate operation. Through the principle of association there would also be a strong tendency towards this same habit, as soon as the mind had resolved on any particular action or line of conduct, even before there was any bodily exertion, or if none were requisite. The habitual and firm closure of the mouth would thus come to show decision of character; and decision readily passes into obstinacy.

CHAPTER X.

HATRED AND ANGER.

Hatred—Rage, effects of on the system—Uncovering of the teeth—Rage in the insane—Anger and indignation—As expressed by the various races of man—Sneering and defiance—The uncovering of the canine tooth on one side of the face.

IF we have suffered or expect to suffer some wilful injury from a man, or if he is in any way offensive to us, we dislike him; and dislike easily rises into hatred. Such feelings, if experienced in a moderate degree, are not clearly expressed by any movement of the body or features, excepting perhaps by a certain gravity of behaviour, or by some ill-temper. Few individuals, however, can long reflect about a hated person, without feeling and exhibiting signs of indignation or rage. But if the offending person be quite insignificant, we experience merely disdain or contempt. If, on the other hand, he is all-powerful, then hatred passes into terror, as when a slave thinks about a cruel master, or a savage about a bloodthirsty malignant deity.[1] Most of our emotions are so closely connected with their expression, that they hardly exist if the body remains passive—the nature of the expression depending in chief part on the

---

[1] See some remarks to this effect by Mr. Bain, 'The Emotions and the Will,' 2nd edit. 1865, p. 127.

nature of the actions which have been habitually performed under this particular state of the mind. A man, for instance, may know that his life is in the extremest peril, and may strongly desire to save it; yet, as Louis XVI. said, when surrounded by a fierce mob, "Am I afraid? feel my pulse." So a man may intensely hate another, but until his bodily frame is affected, he cannot be said to be enraged.

Rage.—I have already had occasion to treat of this emotion in the third chapter, when discussing the direct influence of the excited sensorium on the body, in combination with the effects of habitually associated actions. Rage exhibits itself in the most diversified manner. The heart and circulation are always affected; the face reddens or becomes purple, with the veins on the forehead and neck distended. The reddening of the skin has been observed with the copper-coloured Indians of South America,[2] and even, as it is said, on the white cicatrices left by old wounds on negroes.[3] Monkeys also redden from passion. With one of my own infants, under four months old, I repeatedly observed that the first symptom of an approaching passion was the rushing of the blood into his bare scalp. On the other hand, the action of the heart is sometimes so much impeded by great rage, that the countenance becomes pallid or livid,[4] and not a few men with heart-disease have dropped down dead under this powerful emotion.

---

[2] Rengger, Naturgesch. der Säugethiere von Paraguay, 1830, s. 3.

[3] Sir C. Bell, 'Anatomy of Expression,' p. 96. On the other hand, Dr. Burgess ('Physiology of Blushing,' 1839, p. 31) speaks of the reddening of a cicatrix in a negress as of the nature of a blush.

[4] Moreau and Gratiolet have discussed the colour of the face under the influence of intense passion: see the edit. of 1820 of Lavater, vol. iv. pp. 282 and 300; and Gratiolet, 'De la Physionomie,' p. 345.

The respiration is likewise affected; the chest heaves, and the dilated nostrils quiver.[5] As Tennyson writes, "sharp breaths of anger puffed her fairy nostrils out." Hence we have such expressions as "breathing out vengeance," and "fuming with anger."[6]

The excited brain gives strength to the muscles, and at the same time energy to the will. The body is commonly held erect ready for instant action, but sometimes it is bent forward towards the offending person, with the limbs more or less rigid. The mouth is generally closed with firmness, showing fixed determination, and the teeth are clenched or ground together. Such gestures as the raising of the arms, with the fists clenched, as if to strike the offender, are common. Few men in a great passion, and telling some one to begone, can resist acting as if they intended to strike or push the man violently away. The desire, indeed, to strike often becomes so intolerably strong, that inanimate objects are struck or dashed to the ground; but the gestures frequently become altogether purposeless or frantic. Young children, when in a violent rage roll on the ground on their backs or bellies, screaming, kicking, scratching, or

---

[5] Sir C. Bell ('Anatomy of Expression,' pp. 91, 107) has fully discussed this subject. Moreau remarks (in the edit. of 1820 of 'La Physionomie, par G. Lavater,' vol. iv. p. 237), and quotes Portal in confirmation, that asthmatic patients acquire permanently expanded nostrils, owing to the habitual contraction of the elevatory muscles of the wings of the nose. The explanation by Dr. Piderit ('Mimik und Physiognomik,' s. 82) of the distension of the nostrils, namely, to allow free breathing whilst the mouth is closed and the teeth clenched, does not appear to be nearly so correct as that by Sir C. Bell, who attributes it to the sympathy (*i. e.* habitual co-action) of all the respiratory muscles. The nostrils of an angry man may be seen to become dilated, although his mouth is open.

[6] Mr. Wedgwood, 'On the Origin of Language,' 1866, p. 76. He also observes that the sound of hard breathing "is represented by the syllables *puff*, *huff*, *whiff*, whence *a huff* is a fit of ill-temper."

biting everything within reach. So it is, as I hear from Mr. Scott, with Hindoo children; and, as we have seen, with the young of the anthropomorphous apes.

But the muscular system is often affected in a wholly different way; for trembling is a frequent consequence of extreme rage. The paralysed lips then refuse to obey the will, "and the voice sticks in the throat;"[7] or it is rendered loud, harsh, and discordant. If there be much and rapid speaking, the mouth froths. The hair sometimes bristles; but I shall return to this subject in another chapter, when I treat of the mingled emotions of rage and terror. There is in most cases a strongly-marked frown on the forehead; for this follows from the sense of anything displeasing or difficult, together with concentration of mind. But sometimes the brow, instead of being much contracted and lowered, remains smooth, with the glaring eyes kept widely open. The eyes are always bright, or may, as Homer expresses it, glisten with fire. They are sometimes bloodshot, and are said to protrude from their sockets—the result, no doubt, of the head being gorged with blood, as shown by the veins being distended. According to Gratiolet,[8] the pupils are always contracted in rage, and I hear from Dr. Crichton Browne that this is the case in the fierce delirium of meningitis; but the movements of the iris under the influence of the different emotions is a very obscure subject.

Shakspeare sums up the chief characteristics of rage as follows:—

> "In peace there's nothing so becomes a man,
> As modest stillness and humility;
> But when the blast of war blows in our ears,

---

[7] Sir C. Bell ('Anatomy of Expression,' p. 95) has some excellent remarks on the expression of rage.

[8] 'De la Physionomie,' 1865, p. 346.

Then imitate the action of the tiger:
Stiffen the sinews, summon up the blood,
Then lend the eye a terrible aspect;
Now set the teeth, and stretch the nostril wide,
Hold hard the breath, and bend up every spirit
To his full height! On, on, you noblest English."
Henry V., act iii. sc. 1.

The lips are sometimes protruded during rage in a manner, the meaning of which I do not understand, unless it depends on our descent from some ape-like animal. Instances have been observed, not only with Europeans, but with the Australians and Hindoos. The lips, however, are much more commonly retracted, the grinning or clenched teeth being thus exposed. This has been noticed by almost every one who has written on expression.[9] The appearance is as if the teeth were uncovered, ready for seizing or tearing an enemy, though there may be no intention of acting in this manner. Mr. Dyson Lacy has seen this grinning expression with the Australians, when quarrelling, and so has Gaika with the Kafirs of South America. Dickens,[10] in speaking of an atrocious murderer who had just been caught, and was surrounded by a furious mob, describes "the people as jumping up one behind another, snarling with their teeth, and making at him like wild beasts." Every one who has had much to do with young children must have

[9] Sir C. Bell, 'Anatomy of Expression,' p. 177. Gratiolet (De la Phys. p. 369) says, "les dents se découvrent, et imitent symboliquement l'action de déchirer et de mordre." If, instead of using the vague term *symboliquement*, Gratiolet had said that the action was a remnant of a habit acquired during primeval times when our semi-human progenitors fought together with their teeth, like gorillas and orangs at the present day, he would have been more intelligible. Dr. Piderit ('Mimik,' &c., s. 82) also speaks of the retraction of the upper lip during rage. In an engraving of one of Hogarth's wonderful pictures, passion is represented in the plainest manner by the open glaring eyes, frowning forehead, and exposed grinning teeth.

[10] 'Oliver Twist,' vol. iii. p. 245.

seen how naturally they take to biting, when in a passion. It seems as instinctive in them as in young crocodiles, who snap their little jaws as soon as they emerge from the egg.

A grinning expression and the protrusion of the lips appear sometimes to go together. A close observer says that he has seen many instances of intense hatred (which can hardly be distinguished from rage, more or less suppressed) in Orientals, and once in an elderly English woman. In all these cases there "was a grin, not a scowl —the lips lengthening, the cheeks settling downwards, the eyes half-closed, whilst the brow remained perfectly calm."[11]

This retraction of the lips and uncovering of the teeth during paroxysms of rage, as if to bite the offender, is so remarkable, considering how seldom the teeth are used by men in fighting, that I inquired from Dr. J. Crichton Browne whether the habit was common in the insane whose passions are unbridled. He informs me that he has repeatedly observed it both with the insane and idiotic, and has given me the following illustrations:—

Shortly before receiving my letter, he witnessed an uncontrollable outbreak of anger and delusive jealousy in an insane lady. At first she vituperated her husband, and whilst doing so foamed at the mouth. Next she approached close to him with compressed lips, and a virulent set frown. Then she drew back her lips, especially the corners of the upper lip, and showed her teeth, at the same time aiming a vicious blow at him. A second case is that of an old soldier, who, when he is requested to conform to the rules of the establishment, gives way to discontent, terminating in fury. He commonly begins

---

[11] 'The Spectator,' July 11, 1868, p. 819.

by asking Dr. Browne whether he is not ashamed to treat him in such a manner. He then swears and blasphemes, paces up and down, tosses his arms wildly about, and menaces any one near him. At last, as his exasperation culminates, he rushes up towards Dr. Browne with a peculiar sidelong movement, shaking his doubled fist, and threatening destruction. Then his upper lip may be seen to be raised, especially at the corners, so that his huge canine teeth are exhibited. He hisses forth his curses through his set teeth, and his whole expression assumes the character of extreme ferocity. A similar description is applicable to another man, excepting that he generally foams at the mouth and spits, dancing and jumping about in a strange rapid manner, shrieking out his maledictions in a shrill falsetto voice.

Dr. Browne also informs me of the case of an epileptic idiot, incapable of independent movements, and who spends the whole day in playing with some toys; but his temper is morose and easily roused into fierceness. When any one touches his toys, he slowly raises his head from its habitual downward position, and fixes his eyes on the offender, with a tardy yet angry scowl. If the annoyance be repeated, he draws back his thick lips and reveals a prominent row of hideous fangs (large canines being especially noticeable), and then makes a quick and cruel clutch with his open hand at the offending person. The rapidity of this clutch, as Dr. Browne remarks, is marvellous in a being ordinarily so torpid that he takes about fifteen seconds, when attracted by any noise, to turn his head from one side to the other. If, when thus incensed, a handkerchief, book, or other article, be placed into his hands, he drags it to his mouth and bites it. Mr. Nicol has likewise described to me two cases of insane patients, whose lips are retracted during paroxysms of rage.

Dr. Maudsley, after detailing various strange animal-like traits in idiots, asks whether these are not due to the reappearance of primitive instincts—"a faint echo from a far-distant past, testifying to a kinship which man has almost outgrown." He adds, that as every human brain passes, in the course of its development, through the same stages as those occurring in the lower vertebrate animals, and as the brain of an idiot is in an arrested condition, we may presume that it "will manifest its most primitive functions, and no higher functions." Dr. Maudsley thinks that the same view may be extended to the brain in its degenerated condition in some insane patients; and asks, whence come "the savage snarl, the destructive disposition, the obscene language, the wild howl, the offensive habits, displayed by some of the insane? Why should a human being, deprived of his reason, ever become so brutal in character, as some do, unless he has the brute nature within him?"[12] This question must, as it would appear, be answered in the affirmative.

Anger, Indignation.—These states of the mind differ from rage only in degree, and there is no marked distinction in their characteristic signs. Under moderate anger the action of the heart is a little increased, the colour heightened, and the eyes become bright. The respiration is likewise a little hurried; and as all the muscles serving for this function act in association, the wings of the nostrils are somewhat raised to allow of a free indraught of air; and this is a highly characteristic sign of indignation. The mouth is commonly compressed, and there is almost always a frown on the brow. Instead of the frantic gestures of extreme rage, an indignant man unconsciously throws himself into an atti-

[12] 'Body and Mind,' 1870, pp. 51–53.

tude ready for attacking or striking his enemy, whom he will perhaps scan from head to foot in defiance. He carries his head erect, with his chest well expanded, and the feet planted firmly on the ground. He holds his arms in various positions, with one or both elbows squared, or with the arms rigidly suspended by his sides. With Europeans the fists are commonly clenched.[13] The figures 1 and 2 in Plate VI. are fairly good representations of men simulating indignation. Any one may see in a mirror, if he will vividly imagine that he has been insulted and demands an explanation in an angry tone of voice, that he suddenly and unconsciously throws himself into some such attitude.

Rage, anger, and indignation are exhibited in nearly the same manner throughout the world; and the following descriptions may be worth giving as evidence of this, and as illustrations of some of the foregoing remarks. There is, however, an exception with respect to clenching the fists, which seems confined chiefly to the men who fight with their fists. With the Australians only one of my informants has seen the fists clenched. All agree about the body being held erect; and all, with two exceptions, state that the brows are heavily contracted. Some of them allude to the firmly-compressed mouth, the distended nostrils, and flashing eyes. According to the Rev. Mr. Taplin, rage, with the Australians, is expressed by the lips being protruded, the eyes being widely open; and in the case of the women by their dancing about and casting dust into the air. Another ob-

[13] Le Brun, in his well-known 'Conférence sur l'Expression' ('La Physionomie, par Lavater,' edit. of 1820, vol. ix. p. 268), remarks that anger is expressed by the clenching of the fists. See, to the same effect, Huschke, 'Mimices et Physiognomices, Fragmentum Physiologicum,' 1824, p. 20. Also Sir C. Bell, 'Anatomy of Expression,' p. 219.

server speaks of the native men, when enraged, throwing their arms wildly about.

I have received similar accounts, except as to the clenching of the fists, in regard to the Malays of the Malacca peninsula, the Abyssinians, and the natives of South Africa. So it is with the Dakota Indians of North America; and, according to Mr. Matthews, they then hold their heads erect, frown, and often stalk away with long strides. Mr. Bridges states that the Fuegians, when enraged, frequently stamp on the ground, walk distractedly about, sometimes cry and grow pale. The Rev. Mr. Stack watched a New Zealand man and woman quarrelling, and made the following entry in his note-book: "Eyes dilated, body swayed violently backwards and forwards, head inclined forwards, fists clenched, now thrown behind the body, now directed towards each other's faces." Mr. Swinhoe says that my description agrees with what he has seen of the Chinese, excepting that an angry man generally inclines his body towards his antagonist, and pointing at him, pours forth a volley of abuse.

Lastly, with respect to the natives of India, Mr. J. Scott has sent me a full description of their gestures and expression when enraged. Two low-caste Bengalees disputed about a loan. At first they were calm, but soon grew furious and poured forth the grossest abuse on each other's relations and progenitors for many generations past. Their gestures were very different from those of Europeans; for though their chests were expanded and shoulders squared, their arms remained rigidly suspended, with the elbows turned inwards and the hands alternately clenched and opened. Their shoulders were often raised high, and then again lowered. They looked fiercely at each other from under their lowered and strongly wrinkled brows, and their protruded lips were firmly closed. They approached each other, with heads and necks stretched forwards, and pushed, scratched, and grasped at each other. This protrusion of the head and body seems a common gesture with the enraged; and I have noticed it with degraded English women whilst quarrelling violently in the streets. In such cases it may be presumed that neither party expects to receive a blow from the other.

A Bengalee employed in the Botanic Gardens was accused, in the presence of Mr. Scott, by the native overseer of having stolen a valuable plant. He listened silently and scornfully to the accusation; his attitude erect, chest expanded, mouth closed, lips protruding, eyes firmly set and penetrating. He then defiantly maintained his innocence, with upraised and clenched hands, his head being now pushed forwards, with the eyes widely open and eyebrows raised. Mr. Scott also watched two Mechis, in Sikhim, quarrelling about their share of payment. They soon got into a furious passion, and then their bodies became less erect, with their heads pushed forwards; they made grimaces at each other; their shoulders were raised; their arms rigidly bent inwards at the elbows, and their hands spasmodically closed, but not properly clenched. They continually approached and retreated from each other, and often raised their arms as if to strike, but their hands were open, and no blow was given. Mr. Scott made similar observations on the Lepchas whom he often saw quarrelling, and he noticed that they kept their arms rigid and almost parallel to their bodies, with the hands pushed somewhat backwards and partially closed, but not clenched.

Sneering, Defiance: Uncovering the canine tooth on one side.—The expression which I wish here to consider differs but little from that already described, when the lips are retracted and the grinning teeth exposed. The difference consists solely in the upper lip being retracted in such a manner that the canine tooth on one side of the face alone is shown; the face itself being generally a little upturned and half averted from the person causing offence. The other signs of rage are not necessarily present. This expression may occasionally be observed in a person who sneers at or defies another, though there may be no real anger; as when any one is playfully accused of some fault, and answers, "I scorn the imputation." The expression is not a common one, but I have seen it exhibited with perfect distinctness by a lady who was being quizzed by another person. It was described by Parsons as long ago as 1746, with an engraving, showing the uncovered canine on one side.[14] Mr. Rejlander, without my having made any allusion to the subject, asked me whether I had ever noticed this expression, as he had been much struck by it. He has photographed for me (Plate IV. fig 1) a lady, who sometimes unintentionally displays the canine on one side, and who can do so voluntarily with unusual distinctness.

The expression of a half-playful sneer graduates into one of great ferocity when, together with a heavily frowning brow and fierce eye, the canine tooth is exposed. A Bengalee boy was accused before Mr. Scott of some misdeed. The delinquent did not dare to give vent to his wrath in words, but it was plainly shown on his countenance, sometimes by a defiant frown, and sometimes "by a thoroughly canine snarl." When this was exhibited, "the corner of the lip over the eye-tooth, which happened in this case to be large and projecting, was raised on the side of his accuser, a strong frown

---

[14] Transact. Philosoph. Soc., Appendix, 1746, p. 65.

1

2

being still retained on the brow." Sir C. Bell states[15] that the actor Cooke could express the most determined hate "when with the oblique cast of his eyes he drew up the outer part of the upper lip, and discovered a sharp angular tooth."

The uncovering of the canine tooth is the result of a double movement. The angle or corner of the mouth is drawn a little backwards, and at the same time a muscle which runs parallel to and near the nose draws up the outer part of the upper lip, and exposes the canine on this side of the face. The contraction of this muscle makes a distinct furrow on the cheek, and produces strong wrinkles under the eye, especially at its inner corner. The action is the same as that of a snarling dog; and a dog when pretending to fight often draws up the lip on one side alone, namely that facing his antagonist. Our word *sneer* is in fact the same as *snarl*, which was originally *snar*, the *l* "being merely an element implying continuance of action."[16]

I suspect that we see a trace of this same expression in what is called a derisive or sardonic smile. The lips are then kept joined or almost joined, but one corner of the mouth is retracted on the side towards the derided person; and this drawing back of the corner is part of a true sneer. Although some persons smile more on one side of their face than on the other, it is not easy to understand why in cases of derision the smile, if a real one, should so commonly be confined to one side. I have also on these occasions noticed a slight twitching of the muscle which draws up the outer part

---

[15] 'Anatomy of Expression,' p. 136. Sir C. Bell calls (p. 131) the muscles which uncover the canines the *snarling muscles*.

[16] Hensleigh Wedgwood, 'Dictionary of English Etymology,' 1865, vol. iii. pp. 240, 243.

17

of the upper lip; and this movement, if fully carried out, would have uncovered the canine, and would have produced a true sneer.

Mr. Bulmer, an Australian missionary in a remote part of Gipps' Land, says, in answer to my query about the uncovering of the canine on one side, "I find that the natives in snarling at each other speak with the teeth closed, the upper lip drawn to one side, and a general angry expression of face; but they look direct at the person addressed." Three other observers in Australia, one in Abyssinia, and one in China, answer my query on this head in the affirmative; but as the expression is rare, and as they enter into no details, I am afraid of implicitly trusting them. It is, however, by no means improbable that this animal-like expression may be more common with savages than with civilized races. Mr. Geach is an observer who may be fully trusted, and he has observed it on one occasion in a Malay in the interior of Malacca. The Rev. S. O. Glenie answers, "We have observed this expression with the natives of Ceylon, but not often." Lastly, in North America, Dr. Rothrock has seen it with some wild Indians, and often in a tribe adjoining the Atnahs.

Although the upper lip is certainly sometimes raised on one side alone in sneering at or defying any one, I do not know that this is always the case, for the face is commonly half averted, and the expression is often momentary. The movement being confined to one side may not be an essential part of the expression, but may depend on the proper muscles being incapable of movement excepting on one side. I asked four persons to endeavour to act voluntarily in this manner; two could expose the canine only on the left side, one only on the right side, and the fourth on neither side. Nevertheless it is by no means certain that these same persons, if defying any one in earnest, would not unconsciously have uncovered their canine tooth on the side, whichever it might be, towards the offender. For we have seen that some persons cannot voluntarily make their eyebrows oblique, yet instantly act in this manner when affected by any real, although most trifling, cause of distress. The power of voluntarily uncovering the canine on one side of the face being thus often wholly lost, indicates that it is a rarely used and almost abortive action. It is indeed a surprising fact that man should possess the power, or should exhibit any tendency to its use; for Mr. Sutton has never noticed a snarling action in our nearest allies, namely, the monkeys in the Zoological Gardens, and he is positive that the baboons, though furnished with great canines, never act thus, but uncover all their teeth when feeling savage and ready for an attack. Whether the adult anthropomorphous apes, in the males of whom the canines are much larger than in the females, uncover them when prepared to fight, is not known.

The expression here considered, whether that of a playful sneer or ferocious snarl, is one of the most curious which occurs in man. It reveals his animal descent; for no one, even if rolling on the ground in a deadly grapple with an enemy, and attempting to bite him, would try to use his canine teeth more than his other teeth. We may readily believe from our affinity to the anthropomorphous apes that our male semi-human progenitors possessed great canine teeth, and men are now occasionally born having them of unusually large size, with interspaces in the opposite jaw for their reception.[17] We may further suspect, notwithstanding that we have no support from analogy, that our semi-human progenitors un-

[17] 'The Descent of Man,' 1871, vol. i. p. 126.

covered their canine teeth when prepared for battle, as we still do when feeling ferocious, or when merely sneering at or defying some one, without any intention of making a real attack with our teeth.

CHAPTER XI.

DISDAIN—CONTEMPT—DISGUST—GUILT—PRIDE, ETC. —HELPLESSNESS—PATIENCE—AFFIRMATION AND NEGATION.

Contempt, scorn and disdain, variously expressed—Derisive smile—Gestures expressive of contempt—Disgust —Guilt, deceit, pride, &c.—Helplessness or impotence —Patience—Obstinacy—Shrugging the shoulders common to most of the races of man—Signs of affirmation and negation.

SCORN and disdain can hardly be distinguished from contempt, excepting that they imply a rather more angry frame of mind. Nor can they be clearly distinguished from the feelings discussed in the last chapter under the terms of sneering and defiance. Disgust is a sensation rather more distinct in its nature, and refers to something revolting, primarily in relation to the sense of taste, as actually perceived or vividly imagined; and secondarily to anything which causes a similar feeling, through the sense of smell, touch, and even of eyesight. Nevertheless, extreme contempt, or as it is often called loathing contempt, hardly differs from disgust. These several conditions of the mind are, therefore, nearly related; and each of them may be exhibited in many different ways. Some writers have insisted chiefly on one mode of expression, and others on a different mode.

From this circumstance M. Lemoine has argued [1] that their descriptions are not trustworthy. But we shall immediately see that it is natural that the feelings which we have here to consider should be expressed in many different ways, inasmuch as various habitual actions serve equally well, through the principle of association, for their expression.

Scorn and disdain, as well as sneering and defiance, may be displayed by a slight uncovering of the canine tooth on one side of the face; and this movement appears to graduate into one closely like a smile. Or the smile or laugh may be real, although one of derision; and this implies that the offender is so insignificant that he excites only amusement; but the amusement is generally a pretence. Gaika in his answers to my queries remarks, that contempt is commonly shown by his countrymen, the Kafirs, by smiling; and the Rajah Brooke makes the same observation with respect to the Dyaks of Borneo. As laughter is primarily the expression of simple joy, very young children do not, I believe, ever laugh in derision.

The partial closure of the eyelids, as Duchenne [2] insists, or the turning away of the eyes or of the whole body, are likewise highly expressive of disdain. These actions seem to declare that the despised person is not worth looking at, or is disagreeable to behold. The accompanying photograph (Plate V. fig. 1) by Mr. Rejlander, shows this form of disdain. It represents a young lady, who is supposed to be tearing up the photograph of a despised lover.

The most common method of expressing contempt is

---

[1] 'De la Physionomie et la Parole,' 1865, p. 89.

[2] 'Physionomie Humaine,' Album, Légende viii. p. 35. Gratiolet also speaks (De la Phys. 1865, p. 52) of the turning away of the eyes and body.

1

2

3

by movements about the nose, or round the mouth; but the latter movements, when strongly pronounced, indicate disgust. The nose may be slightly turned up, which apparently follows from the turning up of the upper lip; or the movement may be abbreviated into the mere wrinkling of the nose. The nose is often slightly contracted, so as partly to close the passage;[3] and this is commonly accompanied by a slight snort or expiration. All these actions are the same with those which we employ when we perceive an offensive odour, and wish to exclude or expel it. In extreme cases, as Dr. Piderit remarks,[4] we protrude and raise both lips, or the upper lip alone, so as to close the nostrils as by a valve, the nose being thus turned up. We seem thus to say to the despised person that he smells offensively,[5] in nearly the same manner as we express to him by half-closing our eyelids, or turning away our faces, that he is not worth looking at. It must not, however, be supposed that such ideas actually pass through the mind when we exhibit our contempt; but as whenever we have perceived a dis-

---

[3] Dr. W. Ogle, in an interesting paper on the Sense of Smell ('Medico-Chirurgical Transactions,' vol. liii. p. 268), shows that when we wish to smell carefully, instead of taking one deep nasal inspiration, we draw in the air by a succession of rapid short sniffs. If "the nostrils be watched during this process, it will be seen that, so far from dilating, they actually contract at each sniff. The contraction does not include the whole anterior opening, but only the posterior portion." He then explains the cause of this movement. When, on the other hand, we wish to exclude any odour, the contraction, I presume, affects only the anterior part of the nostrils.

[4] 'Mimik und Physiognomik,' ss. 84, 93. Gratiolet (ibid. p. 155) takes nearly the same view with Dr. Piderit respecting the expression of contempt and disgust.

[5] Scorn implies a strong form of contempt; and one of the roots of the word 'scorn' means, according to Mr. Wedgwood (Dict. of English Etymology, vol. iii. p. 125), ordure or dirt. A person who is scorned is treated like dirt.

agreeable odour or seen a disagreeable sight, actions of this kind have been performed, they have become habitual or fixed, and are now employed under any analogous state of mind.

Various odd little gestures likewise indicate contempt; for instance, *snapping one's fingers.* This, as Mr. Tylor remarks,[6] "is not very intelligible as we generally see it; but when we notice that the same sign made quite gently, as if rolling some tiny object away between the finger and thumb, or the sign of flipping it away with the thumb-nail and forefinger, are usual and well-understood deaf-and-dumb gestures, denoting anything tiny, insignificant, contemptible, it seems as though we had exaggerated and conventionalized a perfectly natural action, so as to lose sight of its original meaning. There is a curious mention of this gesture by Strabo." Mr. Washington Matthews informs me that, with the Dakota Indians of North America, contempt is shown not only by movements of the face, such as those above described, but "conventionally, by the hand being closed and held near the breast, then, as the forearm is suddenly extended, the hand is opened and the fingers separated from each other. If the person at whose expense the sign is made is present, the hand is moved towards him, and the head sometimes averted from him." This sudden extension and opening of the hand perhaps indicates the dropping or throwing away a valueless object.

The term 'disgust,' in its simplest sense, means something offensive to the taste. It is curious how readily this feeling is excited by anything unusual in the appearance, odour, or nature of our food. In Tierra del Fuego a native touched with his finger some cold pre-

---

[6] 'Early History of Mankind,' 2nd edit. 1870, p. 45.

served meat which I was eating at our bivouac, and plainly showed utter disgust at its softness; whilst I felt utter disgust at my food being touched by a naked savage, though his hands did not appear dirty. A smear of soup on a man's beard looks disgusting, though there is of course nothing disgusting in the soup itself. I presume that this follows from the strong association in our minds between the sight of food, however circumstanced, and the idea of eating it.

As the sensation of disgust primarily arises in connection with the act of eating or tasting, it is natural that its expression should consist chiefly in movements round the mouth. But as disgust also causes annoyance, it is generally accompanied by a frown, and often by gestures as if to push away or to guard oneself against the offensive object. In the two photographs (figs. 2 and 3, on Plate V.) Mr. Rejlander has simulated this expression with some success. With respect to the face, moderate disgust is exhibited in various ways; by the mouth being widely opened, as if to let an offensive morsel drop out; by spitting; by blowing out of the protruded lips; or by a sound as of clearing the throat. Such guttural sounds are written *ach* or *ugh*; and their utterance is sometimes accompanied by a shudder, the arms being pressed close to the sides and the shoulders raised in the same manner as when horror is experienced.[7] Extreme disgust is expressed by movements round the mouth identical with those preparatory to the act of vomiting. The mouth is opened widely, with the upper lip strongly retracted, which wrinkles the sides of the nose, and with the lower lip protruded and everted as much as possible. This latter movement requires the

[7] See, to this effect, Mr. Hensleigh Wedgwood's Introduction to the 'Dictionary of English Etymology,' 2nd edit. 1872, p. xxxvii.

contraction of the muscles which draw downwards the corners of the mouth.[8]

It is remarkable how readily and instantly retching or actual vomiting is induced in some persons by the mere idea of having partaken of any unusual food, as of an animal which is not commonly eaten; although there is nothing in such food to cause the stomach to reject it. When vomiting results, as a reflex action, from some real cause—as from too rich food, or tainted meat, or from an emetic—it does not ensue immediately, but generally after a considerable interval of time. Therefore, to account for retching or vomiting being so quickly and easily excited by a mere idea, the suspicion arises that our progenitors must formerly have had the power (like that possessed by ruminants and some other animals) of voluntarily rejecting food which disagreed with them, or which they thought would disagree with them; and now, though this power has been lost, as far as the will is concerned, it is called into involuntary action, through the force of a formerly well-established habit, whenever the mind revolts at the idea of having partaken of any kind of food, or at anything disgusting. This suspicion receives support from the fact, of which I am assured by Mr. Sutton, that the monkeys in the Zoological Gardens often vomit whilst in perfect health, which looks as if the act were voluntary. We can see that as man is able to communicate by language to his children and others, the knowledge of the kinds of food to be avoided, he would have little occasion to use the faculty of voluntary rejection; so that this power would tend to be lost through disuse.

[8] Duchenne believes that in the eversion of the lower lip, the corners are drawn downwards by the *depressores anguli oris*. Henle (Handbuch d. Anat. des Menschen, 1858, B. i. s. 151) concludes that this is effected by the *musculus quadratus menti*.

As the sense of smell is so intimately connected with that of taste, it is not surprising that an excessively bad odour should excite retching or vomiting in some persons, quite as readily as the thought of revolting food does; and that, as a further consequence, a moderately offensive odour should cause the various expressive movements of disgust. The tendency to retch from a fetid odour is immediately strengthened in a curious manner by some degree of habit, though soon lost by longer familiarity with the cause of offence and by voluntary restraint. For instance, I wished to clean the skeleton of a bird, which had not been sufficiently macerated, and the smell made my servant and myself (we not having had much experience in such work) retch so violently, that we were compelled to desist. During the previous days I had examined some other skeletons, which smelt slightly; yet the odour did not in the least affect me, but, subsequently for several days, whenever I handled these same skeletons, they made me retch.

From the answers received from my correspondents it appears that the various movements, which have now been described as expressing contempt and disgust, prevail throughout a large part of the world. Dr. Rothrock, for instance, answers with a decided affirmative with respect to certain wild Indian tribes of North America. Crantz says that when a Greenlander denies anything with contempt or horror he turns up his nose, and gives a slight sound through it.[9] Mr. Scott has sent me a graphic description of the face of a young Hindoo at the sight of castor-oil, which he was compelled occasionally to take. Mr. Scott has also seen the same expression on the faces of high-caste natives who have

---

[9] As quoted by Tylor, 'Primitive Culture,' 1871, vol. i. p. 169.

approached close to some defiling object. Mr. Bridges says that the Fuegians "express contempt by shooting out the lips and hissing through them, and by turning up the nose." The tendency either to snort through the nose, or to make a noise expressed by *ugh* or *ach*, is noticed by several of my correspondents.

Spitting seems an almost universal sign of contempt or disgust; and spitting obviously represents the rejection of anything offensive from the mouth. Shakspeare makes the Duke of Norfolk say, "I spit at him—call him a slanderous coward and a villain." So, again, Falstaff says, "Tell thee what, Hal,—if I tell thee a lie, spit in my face." Leichhardt remarks that the Australians "interrupted their speeches by spitting, and uttering a noise like pooh! pooh! apparently expressive of their disgust." And Captain Burton speaks of certain negroes "spitting with disgust upon the ground." [10] Captain Speedy informs me that this is likewise the case with the Abyssinians. Mr. Geach says that with the Malays of Malacca the expression of disgust "answers to spitting from the mouth;" and with the Fuegians, according to Mr. Bridges "to spit at one is the highest mark of contempt."

I never saw disgust more plainly expressed than on the face of one of my infants at the age of five months, when, for the first time, some cold water, and again a month afterwards, when a piece of ripe cherry was put into his mouth. This was shown by the lips and whole mouth assuming a shape which allowed the contents to run or fall quickly out; the tongue being likewise protruded. These movements were accompanied by a little shudder. It was all the more comical, as I doubt whether

---

[10] Both these quotations are given by Mr. H. Wedgwood, 'On the Origin of Language,' 1866, p. 75.

the child felt real disgust—the eyes and forehead expressing much surprise and consideration. The protrusion of the tongue in letting a nasty object fall out of the mouth, may explain how it is that lolling out the tongue universally serves as a sign of contempt and hatred.[11]

We have now seen that scorn, disdain, contempt, and disgust are expressed in many different ways, by movements of the features, and by various gestures; and that these are the same throughout the world. They all consist of actions representing the rejection or exclusion of some real object which we dislike or abhor, but which does not excite in us certain other strong emotions, such as rage or terror; and through the force of habit and association similar actions are performed, whenever any analogous sensation arises in our minds.

Jealousy, Envy, Avarice, Revenge, Suspicion, Deceit, Slyness, Guilt, Vanity, Conceit, Ambition, Pride, Humility, &c.—It is doubtful whether the greater number of the above complex states of mind are revealed by any fixed expression, sufficiently distinct to be described or delineated. When Shakspeare speaks of Envy as *lean-faced*, or *black*, or *pale*, and Jealousy as "*the green-eyed monster;*" and when Spenser describes Suspicion as "*foul, ill-favoured, and grim,*" they must have felt this difficulty. Nevertheless, the above feelings—at least many of them—can be detected by the eye; for instance, conceit; but we are often guided in a much greater degree than we suppose by our previous knowledge of the persons or circumstances.

My correspondents almost unanimously answer in the affirmative to my query, whether the expression of

---

[11] This is stated to be the case by Mr. Tylor (Early Hist. of Mankind, 2nd edit. 1870, p. 52); and he adds, "it is not clear why this should be so."

guilt and deceit can be recognized amongst the various races of man; and I have confidence in their answers, as they generally deny that jealousy can thus be recognized. In the cases in which details are given, the eyes are almost always referred to. The guilty man is said to avoid looking at his accuser, or to give him stolen looks. The eyes are said "to be turned askant," or "to waver from side to side," or "the eyelids to be lowered and partly closed." This latter remark is made by Mr. Hagenauer with respect to the Australians, and by Gaika with respect to the Kafirs. The restless movements of the eyes apparently follow, as will be explained when we treat of blushing, from the guilty man not enduring to meet the gaze of his accuser. I may add, that I have observed a guilty expression, without a shade of fear, in some of my own children at a very early age. In one instance the expression was unmistakably clear in a child two years and seven months old, and led to the detection of his little crime. It was shown, as I record in my notes made at the time, by an unnatural brightness in the eyes, and by an odd, affected manner, impossible to describe.

Slyness is also, I believe, exhibited chiefly by movements about the eyes; for these are less under the control of the will, owing to the force of long-continued habit, than are the movements of the body. Mr. Herbert Spencer remarks,[12] "When there is a desire to see something on one side of the visual field without being supposed to see it, the tendency is to check the conspicuous movement of the head, and to make the required adjustment entirely with the eyes; which are, therefore, drawn very much to one side. Hence, when the eyes are turned to one side, while the face is not

---

[12] 'Principles of Psychology,' 2nd edit. 1872, p. 552.

turned to the same side, we get the natural language of what is called slyness."

Of all the above-named complex emotions, Pride, perhaps, is the most plainly expressed. A proud man exhibits his sense of superiority over others by holding his head and body erect. He is haughty (*haut*), or high, and makes himself appear as large as possible; so that metaphorically he is said to be swollen or puffed up with pride. A peacock or a turkey-cock strutting about with puffed-up feathers, is sometimes said to be an emblem of pride.[13] The arrogant man looks down on others, and with lowered eyelids hardly condescends to see them; or he may show his contempt by slight movements, such as those before described, about the nostrils or lips. Hence the muscle which everts the lower lip has been called the *musculus superbus*. In some photographs of patients affected by a monomania of pride, sent me by Dr. Crichton Browne, the head and body were held erect, and the mouth firmly closed. This latter action, expressive of decision, follows, I presume, from the proud man feeling perfect self-confidence in himself. The whole expression of pride stands in direct antithesis to that of humility; so that nothing need here be said of the latter state of mind.

Helplessness, Impotence: Shrugging the shoulders.—When a man wishes to show that he cannot do something, or prevent something being done, he often raises with a quick movement both shoulders. At the same time, if the whole gesture is completed, he bends his elbows closely inwards, raises his open hands, turning

[13] Gratiolet (De la Phys. p. 351) makes this remark, and has some good observations on the expression of pride. See Sir C. Bell ('Anatomy of Expression,' p. 111) on the action of the *musculus superbus*.

them outwards, with the fingers separated. The head is often thrown a little on one side; the eyebrows are elevated, and this causes wrinkles across the forehead. The mouth is generally opened. I may mention, in order to show how unconsciously the features are thus acted on, that though I had often intentionally shrugged my shoulders to observe how my arms were placed, I was not at all aware that my eyebrows were raised and mouth opened, until I looked at myself in a glass; and since then I have noticed the same movements in the faces of others. In the accompanying Plate VI., figs. 3 and 4, Mr. Rejlander has successfully acted the gesture of shrugging the shoulders.

Englishmen are much less demonstrative than the men of most other European nations, and they shrug their shoulders far less frequently and energetically than Frenchmen or Italians do. The gesture varies in all degrees from the complex movement, just described, to only a momentary and scarcely perceptible raising of both shoulders; or, as I have noticed in a lady sitting in an arm-chair, to the mere turning slightly outwards of the open hands with separated fingers. I have never seen very young English children shrug their shoulders, but the following case was observed with care by a medical professor and excellent observer, and has been communicated to me by him. The father of this gentleman was a Parisian, and his mother a Scotch lady. His wife is of British extraction on both sides, and my informant does not believe that she ever shrugged her shoulders in her life. His children have been reared in England, and the nursemaid is a thorough Englishwoman, who has never been seen to shrug her shoulders. Now, his eldest daughter was observed to shrug her shoulders at the age of between sixteen and eighteen months; her mother exclaiming at the time, "Look at

2

1

4

3

the little French girl shrugging her shoulders!" At first she often acted thus, sometimes throwing her head a little backwards and on one side, but she did not, as far as was observed, move her elbows and hands in the usual manner. The habit gradually wore away, and now, when she is a little over four years old, she is never seen to act thus. The father is told that he sometimes shrugs his shoulders, especially when arguing with any one; but it is extremely improbable that his daughter should have imitated him at so early an age; for, as he remarks, she could not possibly have often seen this gesture in him. Moreover, if the habit had been acquired through imitation, it is not probable that it would so soon have been spontaneously discontinued by this child, and, as we shall immediately see, by a second child, though the father still lived with his family. This little girl, it may be added, resembles her Parisian grandfather in countenance to an almost absurd degree. She also presents another and very curious resemblance to him, namely, by practising a singular trick. When she impatiently wants something, she holds out her little hand, and rapidly rubs the thumb against the index and middle finger: now this same trick was frequently performed under the same circumstances by her grandfather.

This gentleman's second daughter also shrugged her shoulders before the age of eighteen months, and afterwards discontinued the habit. It is of course possible that she may have imitated her elder sister; but she continued it after her sister had lost the habit. She at first resembled her Parisian grandfather in a less degree than did her sister at the same age, but now in a greater degree. She likewise practises to the present time the peculiar habit of rubbing together, when impatient, her thumb and two of her fore-fingers.

In this latter case we have a good instance, like those given in a former chapter, of the inheritance of a trick or gesture; for no one, I presume, will attribute to mere coincidence so peculiar a habit as this, which was common to the grandfather and his two grandchildren who had never seen him.

Considering all the circumstances with reference to these children shrugging their shoulders, it can hardly be doubted that they have inherited the habit from their French progenitors, although they have only one quarter French blood in their veins, and although their grandfather did not often shrug his shoulders. There is nothing very unusual, though the fact is interesting, in these children having gained by inheritance a habit during early youth, and then discontinuing it; for it is of frequent occurrence with many kinds of animals that certain characters are retained for a period by the young, and are then lost.

As it appeared to me at one time improbable in a high degree that so complex a gesture as shrugging the shoulders, together with the accompanying movements, should be innate, I was anxious to ascertain whether the blind and deaf Laura Bridgman, who could not have learnt the habit by imitation, practised it. And I have heard, through Dr. Innes, from a lady who has lately had charge of her, that she does shrug her shoulders, turn in her elbows, and raise her eyebrows in the same manner as other people, and under the same circumstances. I was also anxious to learn whether this gesture was practised by the various races of man, especially by those who never have had much intercourse with Europeans. We shall see that they act in this manner; but it appears that the gesture is sometimes confined to merely raising or shrugging the shoulders, without the other movements.

Mr. Scott has frequently seen this gesture in the Bengalees and Dhangars (the latter constituting a distinct race) who are employed in the Botanic Garden at Calcutta; when, for instance, they have declared that they could not do some work, such as lifting a heavy weight. He ordered a Bengalee to climb a lofty tree; but the man, with a shrug of his shoulders and a lateral shake of his head, said he could not. Mr. Scott knowing that the man was lazy, thought he could, and insisted on his trying. His face now became pale, his arms dropped to his sides, his mouth and eyes were widely opened, and again surveying the tree, he looked askant at Mr. Scott, shrugged his shoulders, inverted his elbows, extended his open hands, and with a few quick lateral shakes of the head declared his inability. Mr. H. Erskine has likewise seen the natives of India shrugging their shoulders; but he has never seen the elbows turned so much inwards as with us; and whilst shrugging their shoulders they sometimes lay their uncrossed hands on their breasts.

With the wild Malays of the interior of Malacca, and with the Bugis (true Malays, though speaking a different language), Mr. Geach has often seen this gesture. I presume that it is complete, as, in answer to my query descriptive of the movements of the shoulders, arms, hands, and face, Mr. Geach remarks, "it is performed in a beautiful style." I have lost an extract from a scientific voyage, in which shrugging the shoulders by some natives (Micronesians) of the Caroline Archipelago in the Pacific Ocean, was well described. Capt. Speedy informs me that the Abyssinians shrug their shoulders, but enters into no details. Mrs. Asa Gray saw an Arab dragoman in Alexandria acting exactly as described in my query, when an old gentleman, on whom he attended, would not go in the proper direction which had been pointed out to him.

Mr. Washington Matthews says, in reference to the wild Indian tribes of the western parts of the United States, "I have on a few occasions detected men using a slight apologetic shrug, but the rest of the demonstration which you describe I have not witnessed." Fritz Müller informs me that he has seen the negroes in Brazil shrugging their shoulders; but it is of course possible that they may have learnt to do so by imitating the Portuguese. Mrs. Barber has never seen this gesture with the Kafirs of South Africa; and Gaika, judging from his answer, did not even understand what was meant by my description. Mr. Swinhoe is also doubtful about the Chinese; but he has seen them, under the circumstances which would make us shrug our shoulders, press their right elbow against their side, raise their eyebrows, lift up their hand with the palm directed towards the person addressed, and shake it from right to left. Lastly, with respect to the Australians, four of my informants answer by a simple negative, and one by a simple affirmative. Mr. Bunnett, who has had excellent opportunities for observation on the borders of the Colony of Victory, also answers by a "yes," adding that the gesture is performed "in a more subdued and less demonstrative manner than is the case with civilized nations." This circumstance may account for its not having been noticed by four of my informants.

These statements, relating to Europeans, Hindoos, the hill-tribes of India, Malays, Micronesians, Abyssinians, Arabs, Negroes, Indians of North America, and apparently to the Australians—many of these natives having had scarcely any intercourse with Europeans—are sufficient to show that shrugging the shoulders, accompanied in some cases by the other proper movements, is a gesture natural to mankind.

This gesture implies an unintentional or unavoidable action on our own part, or one that we cannot perform; or an action performed by another person which we cannot prevent. It accompanies such speeches as, "It was not my fault;" "It is impossible for me to grant this favour;" "He must follow his own course, I cannot stop him." Shrugging the shoulders likewise expresses patience, or the absence of any intention to resist. Hence the muscles which raise the shoulders are sometimes called, as I have been informed by an artist, "the patience muscles." Shylock the Jew, says,

> "Signor Antonio, many a time and oft
> In the Rialto have you rated me
> About my monies and usances;
> Still have I borne it with a patient shrug."
> *Merchant of Venice*, act i. sc. 3.

Sir C. Bell has given [14] a life-like figure of a man, who is shrinking back from some terrible danger, and is on the point of screaming out in abject terror. He is represented with his shoulders lifted up almost to his ears; and this at once declares that there is no thought of resistance.

As shrugging the shoulders generally implies "I cannot do this or that," so by a slight change, it sometimes implies "I won't do it." The movement then expresses a dogged determination not to act. Olmsted describes [15] an Indian in Texas as giving a great shrug to his shoulders, when he was informed that a party of men were Germans and not Americans, thus expressing that he would have nothing to do with them. Sulky and

---

[14] 'Anatomy of Expression,' p. 166.
[15] 'Journey through Texas,' p. 352.

obstinate children may be seen with both their shoulders raised high up; but this movement is not associated with the others which generally accompany a true shrug. An excellent observer [16] in describing a young man who was determined not to yield to his father's desire, says, "He thrust his hands deep down into his pockets, and set up his shoulders to his ears, which was a good warning that, come right or wrong, this rock should fly from its firm base as soon as Jack would; and that any remonstrance on the subject was purely futile." As soon as the son got his own way, he "put his shoulders into their natural position."

Resignation is sometimes shown by the open hands being placed, one over the other, on the lower part of the body. I should not have thought this little gesture worth even a passing notice, had not Dr. W. Ogle remarked to me that he had two or three times observed it in patients who were preparing for operations under chloroform. They exhibited no great fear, but seemed to declare by this posture of their hands, that they had made up their minds, and were resigned to the inevitable.

We may now inquire why men in all parts of the world when they feel,—whether or not they wish to show this feeling,—that they cannot or will not do something, or will not resist something if done by another, shrug their shoulders, at the same time often bending in their elbows, showing the palms of their hands with extended fingers, often throwing their heads a little on one side, raising their eyebrows, and opening their mouths. These states of the mind are either simply passive, or show a determination not to act. None of the above movements are of the least service. The explanation lies, I

[16] Mrs. Oliphant, 'The Brownlows,' vol. ii. p. 206.

cannot doubt, in the principle of unconscious antithesis. This principle here seems to come into play as clearly as in the case of a dog, who, when feeling savage, puts himself in the proper attitude for attacking and for making himself appear terrible to his enemy; but as soon as he feels affectionate, throws his whole body into a directly opposite attitude, though this is of no direct use to him.

Let it be observed how an indignant man, who resents, and will not submit to some injury, holds his head erect, squares his shoulders, and expands his chest. He often clenches his fists, and puts one or both arms in the proper position for attack or defence, with the muscles of his limbs rigid. He frowns,—that is, he contracts and lowers his brows,—and, being determined, closes his mouth. The actions and attitude of a helpless man are, in every one of these respects, exactly the reverse. In Plate VI. we may imagine one of the figures on the left side to have just said, "What do you mean by insulting me?" and one of the figures on the right side to answer, "I really could not help it." The helpless man unconsciously contracts the muscles of his forehead which are antagonistic to those that cause a frown, and thus raises his eyebrows; at the same time he relaxes the muscles about the mouth, so that the lower jaw drops. The antithesis is complete in every detail, not only in the movements of the features, but in the position of the limbs and in the attitude of the whole body, as may be seen in the accompanying plate. As the helpless or apologetic man often wishes to show his state of mind, he then acts in a conspicuous or demonstrative manner.

In accordance with the fact that squaring the elbows and clenching the fists are gestures by no means universal with the men of all races, when they feel indignant and are prepared to attack their enemy, so it appears that a helpless or apologetic frame of mind is expressed in many parts of the world by merely shrugging the shoulders, without turning inwards the elbows and opening the hands. The man or child who is obstinate, or one who is resigned to some great misfortune, has in neither case any idea of resistance by active means; and he expresses this state of mind, by simply keeping his shoulders raised; or he may possibly fold his arms across his breast.

Signs of affirmation or approval, and of negation or disapproval: nodding and shaking the head.—I was curious to ascertain how far the common signs used by us in affirmation and negation were general throughout the world. These signs are indeed to a certain extent expressive of our feelings, as we give a vertical nod of approval with a smile to our children, when we approve of their conduct; and shake our heads laterally with a frown, when we disapprove. With infants, the first act of denial consists in refusing food; and I repeatedly noticed with my own infants, that they did so by withdrawing their heads laterally from the breast, or from anything offered them in a spoon. In accepting food and taking it into their mouths, they incline their heads forwards. Since making these observations I have been informed that the same idea had occurred to Charma.[17] It deserves notice that in accepting or taking food, there is only a single movement forward, and a single nod implies an affirmation. On the other hand, in refusing food, especially if it be pressed on them, children frequently move their heads several times from side to side, as we do in shaking our heads in negation. Moreover, in the case of refusal, the head is not rarely thrown backwards, or the mouth is closed, so that these movements

[17] 'Essai sur le Langage,' 2nd edit. 1846. I am much indebted to Miss Wedgwood for having given me this information, with an extract from the work.

might likewise come to serve as signs of negation. Mr. Wedgwood remarks on this subject,[18] that "when the voice is exerted with closed teeth or lips, it produces the sound of the letter *n* or *m*. Hence we may account for the use of the particle *ne* to signify negation, and possibly also of the Greek *μή* in the same sense."

That these signs are innate or instinctive, at least with Anglo-Saxons, is rendered highly probable by the blind and deaf Laura Bridgman "constantly accompanying her *yes* with the common affirmative nod, and her *no* with our negative shake of the head." Had not Mr. Lieber stated to the contrary,[19] I should have imagined that these gestures might have been acquired or learnt by her, considering her wonderful sense of touch and appreciation of the movements of others. With microcephalous idiots, who are so degraded that they never learn to speak, one of them is described by Vogt,[20] as answering, when asked whether he wished for more food or drink, by inclining or shaking his head. Schmalz, in his remarkable dissertation on the education of the deaf and dumb, as well as of children raised only one degree above idiotcy, assumes that they can always both make and understand the common signs of affirmation and negation.[21]

Nevertheless if we look to the various races of man, these signs are not so universally employed as I should have expected; yet they seem too general to be ranked as altogether conventional or artificial. My informants assert that both signs are used by the Malays, by the natives of Ceylon, the Chinese, the negroes of the Guinea

---

[18] 'On the Origin of Language,' 1866, p. 91.
[19] 'On the Vocal Sounds of L. Bridgman;' Smithsonian Contributions, 1851, vol. ii. p. 11.
[20] 'Mémoire sur les Microcéphales,' 1867, p. 27.
[21] Quoted by Tylor, 'Early History of Mankind,' 2nd edit. 1870, p. 38.

coast, and, according to Gaika, by the Kafirs of South Africa, though with these latter people Mrs. Barber has never seen a lateral shake used as a negative. With respect to the Australians, seven observers agree that a nod is given in affirmation; five agree about a lateral shake in negation, accompanied or not by some word; but Mr. Dyson Lacy has never seen this latter sign in Queensland, and Mr. Bulmer says that in Gipps' Land a negative is expressed by throwing the head a little backwards and putting out the tongue. At the northern extremity of the continent, near Torres Straits, the natives when uttering a negative "don't shake the head with it, but holding up the right hand, shake it by turning it half round and back again two or three times." [22] The throwing back of the head with a cluck of the tongue is said to be used as a negative by the modern Greeks and Turks, the latter people expressing *yes* by a movement like that made by us when we shake our heads.[23] The Abyssinians, as I am informed by Captain Speedy, express a negative by jerking the head to the right shoulder, together with a slight cluck, the mouth being closed; an affirmation is expressed by the head being thrown backwards and the eyebrows raised for an instant. The Tagals of Luzon, in the Philippine Archipelago, as I hear from Dr. Adolf Meyer, when they say "yes," also throw the head backwards. According to the Rajah Brooke, the Dyaks of Borneo express an affirmation by raising the eyebrows, and a negation by slightly contracting them, together with a peculiar look from the eyes. With the Arabs on the Nile, Professor and Mrs. Asa Gray concluded that nodding in affirmation was rare, whilst

---

[22] Mr. J. B. Jukes, 'Letters and Extracts,' &c. 1871, p. 248.

[23] F. Lieber, 'On the Vocal Sounds,' &c. p. 11. Tylor, ibid. p. 53.

shaking the head in negation was never used, and was not even understood by them. With the Esquimaux [24] a nod means *yes* and a wink *no*. The New Zealanders "elevate the head and chin in place of nodding acquiescence." [25]

With the Hindoos Mr. H. Erskine concludes from inquiries made from experienced Europeans, and from native gentlemen, that the signs of affirmation and negation vary—a nod and a lateral shake being sometimes used as we do; but a negative is more commonly expressed by the head being thrown suddenly backwards and a little to one side, with a cluck of the tongue. What the meaning may be of this cluck of the tongue, which has been observed with various people, I cannot imagine. A native gentleman stated that affirmation is frequently shown by the head being thrown to the left. I asked Mr. Scott to attend particularly to this point, and, after repeated observations, he believes that a vertical nod is not commonly used by the natives in affirmation, but that the head is first thrown backwards either to the left or right, and then jerked obliquely forwards only once. This movement would perhaps have been described by a less careful observer as a lateral shake. He also states that in negation the head is usually held nearly upright, and shaken several times.

Mr. Bridges informs me that the Fuegians nod their heads vertically in affirmation, and shake them laterally in denial. With the wild Indians of North America, according to Mr. Washington Matthews, nodding and shaking the head have been learnt from Europeans, and are not naturally employed. They express affirmation "by describing with the hand (all the fingers except the

---

[24] Dr. King, Edinburgh Phil. Journal, 1845, p. 313.

[25] Tylor, 'Early History of Mankind,' 2nd edit. 1870, p. 53.

index being flexed) a curve downwards and outwards from the body, whilst negation is expressed by moving the open hand outwards, with the palm facing inwards." Other observers state that the sign of affirmation with these Indians is the forefinger being raised, and then lowered and pointed to the ground, or the hand is waved straight forward from the face; and that the sign of negation is the finger or whole hand shaken from side to side.[26] This latter movement probably represents in all cases the lateral shaking of the head. The Italians are said in like manner to move the lifted finger from right to left in negation, as indeed we English sometimes do.

On the whole we find considerable diversity in the signs of affirmation and negation in the different races of man. With respect to negation, if we admit that the shaking of the finger or hand from side to side is symbolic of the lateral movement of the head; and if we admit that the sudden backward movement of the head represents one of the actions often practised by young children in refusing food, then there is much uniformity throughout the world in the signs of negation, and we can see how they originated. The most marked exceptions are presented by the Arabs, Esquimaux, some Australian tribes, and Dyaks. With the latter a frown is the sign of negation, and with us frowning often accompanies a lateral shake of the head.

With respect to nodding in affirmation, the exceptions are rather more numerous, namely with some of the Hindoos, with the Turks, Abyssinians, Dyaks, Tagals, and New Zealanders. The eyebrows are sometimes raised in affirmation, and as a person in bending

---

[26] Lubbock, 'The Origin of Civilization,' 1870, p. 277. Tylor, ibid. p. 38. Lieber (ibid. p. 11) remarks on the negative of the Italians.

his head forwards and downwards naturally looks up to the person whom he addresses, he will be apt to raise his eyebrows, and this sign may thus have arisen as an abbreviation. So again with the New Zealanders, the lifting up the chin and head in affirmation may perhaps represent in an abbreviated form the upward movement of the head after it has been nodded forwards and downwards.

CHAPTER XII.

SURPRISE—ASTONISHMENT—FEAR—HORROR.

Surprise, astonishment—Elevation of the eyebrows—Opening the mouth—Protrusion of the lips—Gestures accompanying surprise—Admiration—Fear—Terror—Erection of the hair—Contraction of the platysma muscle—Dilatation of the pupils—Horror—Conclusion.

ATTENTION, if sudden and close, graduates into surprise; and this into astonishment; and this into stupefied amazement. The latter frame of mind is closely akin to terror. Attention is shown by the eyebrows being slightly raised; and as this state increases into surprise, they are raised to a much greater extent, with the eyes and mouth widely open. The raising of the eyebrows is necessary in order that the eyes should be opened quickly and widely; and this movement produces transverse wrinkles across the forehead. The degree to which the eyes and mouth are opened corresponds with the degree of surprise felt; but these movements must be coordinated; for a widely opened mouth with eyebrows only slightly raised results in a meaningless grimace, as Dr. Duchenne has shown in one of his photographs.[1] On the other hand, a person may often be seen to pretend surprise by merely raising his eyebrows.

Dr. Duchenne has given a photograph of an old man

[1] 'Mécanisme de la Physionomie,' Album, 1862, p. 42.

with his eyebrows well elevated and arched by the galvanization of the frontal muscle; and with his mouth voluntarily opened. This figure expresses surprise with much truth. I showed it to twenty-four persons without a word of explanation, and one alone did not at all understand what was intended. A second person answered terror, which is not far wrong; some of the others, however, added to the words surprise or astonishment, the epithets horrified, woful, painful, or disgusted.

The eyes and mouth being widely open is an expression universally recognized as one of surprise or astonishment. Thus Shakespeare says, "I saw a smith stand with open mouth swallowing a tailor's news." ('King John,' act iv. scene ii.) And again, "They seemed almost, with staring on one another, to tear the cases of their eyes; there was speech in the dumbness, language in their very gesture; they looked as they had heard of a world destroyed." ('Winter's Tale,' act v. scene ii.)

My informants answer with remarkable uniformity to the same effect, with respect to the various races of man; the above movements of the features being often accompanied by certain gestures and sounds, presently to be described. Twelve observers in different parts of Australia agree on this head. Mr. Winwood Reade has observed this expression with the negroes on the Guinea coast. The chief Gaika and others answer *yes* to my query with respect to the Kafirs of South Africa; and so do others emphatically with reference to the Abyssinians, Ceylonese, Chinese, Fuegians, various tribes of North America, and New Zealanders. With the latter, Mr. Stack states that the expression is more plainly shown by certain individuals than by others, though all endeavour as much as possible to conceal their feelings. The Dyaks of Borneo are said by the Rajah Brooke to open their eyes widely, when astonished, often swinging their heads to and fro, and beating their breasts. Mr. Scott informs me that the workmen in the Botanic Gardens at Calcutta are strictly ordered not to smoke; but they often disobey this order, and when suddenly surprised in the act, they first open their eyes and mouths widely. They then often slightly shrug their shoulders, as they perceive that discovery is inevitable, or frown and stamp on the ground from vexation. Soon they recover from their surprise, and abject fear is exhibited by the relaxation of all their muscles; their heads seem to sink between their shoulders; their fallen eyes wander to and fro; and they supplicate forgiveness.

The well-known Australian explorer, Mr. Stuart, has given [2] a striking account of stupefied amazement together with terror in a native who had never before seen a man on horseback. Mr. Stuart approached unseen and called to him from a little distance. "He turned round and saw me. What he imagined I was I do not know; but a finer picture of fear and astonishment I never saw. He stood incapable of moving a limb, riveted to the spot, mouth open and eyes staring. . . . He remained motionless until our black got within a few yards of him, when suddenly throwing down his waddies, he jumped into a mulga bush as high as he could get." He could not speak, and answered not a word to the inquiries made by the black, but, trembling from head to foot, "waved with his hand for us to be off."

That the eyebrows are raised by an innate or instinctive impulse may be inferred from the fact that Laura Bridgman invariably acts thus when astonished, as I have been assured by the lady who has lately had charge of her. As surprise is excited by something unexpected or unknown, we naturally desire, when startled, to per-

[2] 'The Polyglot News Letter,' Melbourne, Dec. 1858, p. 2.

ceive the cause as quickly as possible; and we consequently open our eyes fully, so that the field of vision may be increased, and the eyeballs moved easily in any direction. But this hardly accounts for the eyebrows being so greatly raised as is the case, and for the wild staring of the open eyes. The explanation lies, I believe, in the impossibility of opening the eyes with great rapidity by merely raising the upper lids. To effect this the eyebrows must be lifted energetically. Any one who will try to open his eyes as quickly as possible before a mirror will find that he acts thus; and the energetic lifting up of the eyebrows opens the eyes so widely that they stare, the white being exposed all round the iris. Moreover, the elevation of the eyebrows is an advantage in looking upwards; for as long as they are lowered they impede our vision in this direction. Sir C. Bell gives [3] a curious little proof of the part which the eyebrows play in opening the eyelids. In a stupidly drunken man all the muscles are relaxed, and the eyelids consequently droop, in the same manner as when we are falling asleep. To counteract this tendency the drunkard raises his eyebrows; and this gives to him a puzzled, foolish look, as is well represented in one of Hogarth's drawings. The habit of raising the eyebrows having once been gained in order to see as quickly as possible all around us, the movement would follow from the force of association whenever astonishment was felt from any cause, even from a sudden sound or an idea.

With adult persons, when the eyebrows are raised, the whole forehead becomes much wrinkled in transverse lines; but with children this occurs only to a slight degree. The wrinkles run in lines concentric with each eyebrow, and are partially confluent in the middle.

---

[3] 'The Anatomy of Expression,' p. 106.

They are highly characteristic of the expression of surprise or astonishment. Each eyebrow, when raised, becomes also, as Duchenne remarks,[4] more arched than it was before.

The cause of the mouth being opened when astonishment is felt, is a much more complex affair; and several causes apparently concur in leading to this movement. It has often been supposed [5] that the sense of hearing is thus rendered more acute; but I have watched persons listening intently to a slight noise, the nature and source of which they knew perfectly, and they did not open their mouths. Therefore I at one time imagined that the open mouth might aid in distinguishing the direction whence a sound proceeded, by giving another channel for its entrance into the ear through the eustachian tube, But Dr. W. Ogle [6] has been so kind as to search the best recent authorities on the functions of the eustachian tube, and he informs me that it is almost conclusively proved that it remains closed except during the act of deglutition; and that in persons in whom the tube remains abnormally open, the sense of hearing, as far as external sounds are concerned, is by no means improved; on the contrary, it is impaired by the respiratory sounds being rendered more distinct. If a watch be placed within the mouth, but not allowed to touch the sides, the ticking is heard much less plainly than when held outside. In persons in whom from disease or a cold the eustachian tube is permanently or temporarily closed, the sense of hearing is injured; but this may

---

[4] 'Mécanisme de la Physionomie,' Album, p. 6.

[5] See, for instance, Dr. Piderit ('Mimik und Physiognomik,' s. 88), who has a good discussion on the expression of surprise.

[6] Dr. Murie has also given me information leading to the same conclusion, derived in part from comparative anatomy.

be accounted for by mucus accumulating within the tube, and the consequent exclusion of air. We may therefore infer that the mouth is not kept open under the sense of astonishment for the sake of hearing sounds more distinctly; notwithstanding that most deaf people keep their mouths open.

Every sudden emotion, including astonishment, quickens the action of the heart, and with it the respiration. Now we can breathe, as Gratiolet remarks [7] and as appears to me to be the case, much more quietly through the open mouth than through the nostrils. Therefore, when we wish to listen intently to any sound, we either stop breathing, or breathe as quietly as possible, by opening our mouths, at the same time keeping our bodies motionless. One of my sons was awakened in the night by a noise under circumstances which naturally led to great care, and after a few minutes he perceived that his mouth was widely open. He then became conscious that he had opened it for the sake of breathing as quietly as possible. This view receives support from the reversed case which occurs with dogs. A dog when panting after exercise, or on a hot day, breathes loudly; but if his attention be suddenly aroused, he instantly pricks his ears to listen, shuts his mouth, and breathes quietly, as he is enabled to do, through his nostrils.

When the attention is concentrated for a length of time with fixed earnestness on any object or subject, all the organs of the body are forgotten and neglected; [8] and as the nervous energy of each individual is limited in amount, little is transmitted to any part of the system, excepting that which is at the time brought into energetic action. Therefore many of the muscles tend to

---

[7] 'De la Physionomie,' 1865, p. 234.

[8] See, on this subject, Gratiolet, ibid. p. 254.

become relaxed, and the jaw drops from its own weight. This will account for the dropping of the jaw and open mouth of a man stupefied with amazement, and perhaps when less strongly affected. I have noticed this appearance, as I find recorded in my notes, in very young children when they were only moderately surprised.

There is still another and highly effective cause, leading to the mouth being opened, when we are astonished, and more especially when we are suddenly startled. We can draw a full and deep inspiration much more easily through the widely open mouth than through the nostrils. Now when we start at any sudden sound or sight, almost all the muscles of the body are involuntarily and momentarily thrown into strong action, for the sake of guarding ourselves against or jumping away from the danger, which we habitually associate with anything unexpected. But we always unconsciously prepare ourselves for any great exertion, as formerly explained, by first taking a deep and full inspiration, and we consequently open our mouths. If no exertion follows, and we still remain astonished, we cease for a time to breathe, or breathe as quietly as possible, in order that every sound may be distinctly heard. Or again, if our attention continues long and earnestly absorbed, all our muscles become relaxed, and the jaw, which was at first suddenly opened, remains dropped. Thus several causes concur towards this same movement, whenever surprise, astonishment, or amazement is felt.

Although when thus affected, our mouths are generally opened, yet the lips are often a little protruded. This fact reminds us of the same movement, though in a much more strongly marked degree, in the chimpanzee and orang when astonished. As a strong expiration naturally follows the deep inspiration which accompanies the first sense of startled surprise, and as the lips are often protruded, the various sounds which are then commonly uttered can apparently be accounted for. But sometimes a strong expiration alone is heard; thus Laura Bridgman, when amazed, rounds and protrudes her lips, opens them, and breathes strongly.[9] One of the commonest sounds is a deep *Oh;* and this would naturally follow, as explained by Helmholtz, from the mouth being moderately opened and the lips protruded. On a quiet night some rockets were fired from the 'Beagle,' in a little creek at Tahiti, to amuse the natives; and as each rocket was let off there was absolute silence, but this was invariably followed by a deep groaning *Oh*, resounding all round the bay. Mr. Washington Matthews says that the North American Indians express astonishment by a groan; and the negroes on the West Coast of Africa, according to Mr. Winwood Reade, protrude their lips, and make a sound like *heigh*, *heigh*. If the mouth is not much opened, whilst the lips are considerably protruded, a blowing, hissing, or whistling noise is produced. Mr. R. Brough Smith informs me that an Australian from the interior was taken to the theatre to see an acrobat rapidly turning head over heels: "he was greatly astonished, and protruded his lips, making a noise with his mouth as if blowing out a match." According to Mr. Bulmer the Australians, when surprised, utter the exclamation *korki*, "and to do this the mouth is drawn out as if going to whistle." We Europeans often whistle as a sign of surprise; thus, in a recent novel [10] it is said, "here the man expressed his astonishment and disapprobation by a prolonged whistle." A Kafir girl, as Mr. J. Mansel Weale informs me, "on hearing of the high price of an article, raised her eyebrows and whistled just

[9] Lieber, 'On the Vocal Sounds of Laura Bridgman,' Smithsonian Contributions, 1851, vol. ii. p. 7.
[10] 'Wenderholme,' vol. ii. p. 91.

as a European would." Mr. Wedgwood remarks that such sounds are written down as *whew*, and they serve as interjections for surprise.

According to three other observers, the Australians often evince astonishment by a clucking noise. Europeans also sometimes express gentle surprise by a little clicking noise of nearly the same kind. We have seen that when we are startled, the mouth is suddenly opened; and if the tongue happens to be then pressed closely against the palate, its sudden withdrawal will produce a sound of this kind, which might thus come to express surprise.

Turning to gestures of the body. A surprised person often raises his opened hands high above his head, or by bending his arms only to the level of his face. The flat palms are directed towards the person who causes this feeling, and the straightened fingers are separated. This gesture is represented by Mr. Rejlander in Plate VII. fig. 1. In the 'Last Supper,' by Leonardo da Vinci, two of the Apostles have their hands half uplifted, clearly expressive of their astonishment. A trustworthy observer told me that he had lately met his wife under most unexpected circumstances: "She started, opened her mouth and eyes very widely, and threw up both her arms above her head." Several years ago I was surprised by seeing several of my young children earnestly doing something together on the ground; but the distance was too great for me to ask what they were about. Therefore I threw up my open hands with extended fingers above my head; and as soon as I had done this, I became conscious of the action. I then waited, without saying a word, to see if my children had understood this gesture; and as they came running to me they cried out, "We saw that you were astonished at us." I do not know whether this gesture is common to the various races of man, as I neglected to make inquiries on this head. That it is innate or natural may be inferred from the fact that Laura Bridgman, when amazed, "spreads her arms and turns her hands with extended fingers upwards;"[11] nor is it likely, considering that the feeling of surprise is generally a brief one, that she should have learnt this gesture through her keen sense of touch.

Huschke describes[12] a somewhat different yet allied gesture, which he says is exhibited by persons when astonished. They hold themselves erect, with the features as before described, but with the straightened arms extended backwards—the stretched fingers being separated from each other. I have never myself seen this gesture; but Huschke is probably correct; for a friend asked another man how he would express great astonishment, and he at once threw himself into this attitude.

These gestures are, I believe, explicable on the principle of antithesis. We have seen that an indignant man holds his head erect, squares his shoulders, turns out his elbows, often clenches his fist, frowns, and closes his mouth; whilst the attitude of a helpless man is in every one of these details the reverse. Now, a man in an ordinary frame of mind, doing nothing and thinking of nothing in particular, usually keeps his two arms suspended laxly by his sides, with his hands somewhat flexed, and the fingers near together. Therefore, to raise the arms suddenly, either the whole arms or the fore-arms, to open the palms flat, and to separate the

---

[11] Lieber, 'On the Vocal Sounds,' &c., ibid. p. 7.

[12] Huschke, 'Mimices et Physiognomices,' 1821, p. 18. Gratiolet (De la Phys. p. 255) gives a figure of a man in this attitude, which, however, seems to me expressive of fear combined with astonishment. Le Brun also refers (Lavater, vol. ix. p. 299) to the hands of an astonished man being opened.

fingers,—or, again, to straighten the arms, extending them backwards with separated fingers,—are movements in complete antithesis to those preserved under an indifferent frame of mind, and they are, in consequence, unconsciously assumed by an astonished man. There is, also, often a desire to display surprise in a conspicuous manner, and the above attitudes are well fitted for this purpose. It may be asked why should surprise, and only a few other states of the mind, be exhibited by movements in antithesis to others. But this principle will not be brought into play in the case of those emotions, such as terror, great joy, suffering, or rage, which naturally lead to certain lines of action and produce certain effects on the body, for the whole system is thus preoccupied; and these emotions are already thus expressed with the greatest plainness.

There is another little gesture, expressive of astonishment, of which I can offer no explanation; namely, the hand being placed over the mouth or on some part of the head. This has been observed with so many races of man, that it must have some natural origin. A wild Australian was taken into a large room full of official papers, which surprised him greatly, and he cried out, *cluck*, *cluck*, *cluck*, putting the back of his hand towards his lips. Mrs. Barber says that the Kafirs and Fingoes express astonishment by a serious look and by placing the right hand upon the mouth, uttering the word *mawo*, which means 'wonderful.' The Bushmen are said [13] to put their right hands to their necks, bending their heads backwards. Mr. Winwood Reade has observed that the negroes on the West Coast of Africa, when surprised, clap their hands to their mouths, saying at the same time, "My mouth cleaves to me," *i. e.* to my hands; and

---

[13] Huschke, ibid. p. 18.

he has heard that this is their usual gesture on such occasions. Captain Speedy informs me that the Abyssinians place their right hand to the forehead, with the palm outside. Lastly, Mr. Washington Matthews states that the conventional sign of astonishment with the wild tribes of the western parts of the United States "is made by placing the half-closed hand over the mouth; in doing this, the head is often bent forwards, and words or low groans are sometimes uttered." Catlin [14] makes the same remark about the hand being pressed over the mouth by the Mandans and other Indian tribes.

Admiration.—Little need be said on this head. Admiration apparently consists of surprise associated with some pleasure and a sense of approval. When vividly felt, the eyes are opened and the eyebrows raised; the eyes become bright, instead of remaining blank, as under simple astonishment; and the mouth, instead of gaping open, expands into a smile.

Fear, Terror.—The word 'fear' seems to be derived from what is sudden and dangerous; [15] and that of terror from the trembling of the vocal organs and body. I use the word 'terror' for extreme fear; but some writers think it ought to be confined to cases in which the imagination is more particularly concerned. Fear is often preceded by astonishment, and is so far akin to it, that both lead to the senses of sight and hearing being instantly aroused. In both cases the eyes and mouth are widely opened, and the eyebrows raised. The frightened

---

[14] 'North American Indians,' 3rd edit. 1842, vol. i. p. 105.

[15] H. Wedgwood, Dict. of English Etymology, vol. ii. 1862, p. 35. See, also, Gratiolet ('De la Physionomie,' p. 135) on the sources of such words as 'terror, horror, rigidus, frigidus,' &c.

man at first stands like a statue motionless and breathless, or crouches down as if instinctively to escape observation.

The heart beats quickly and violently, so that it palpitates or knocks against the ribs; but it is very doubtful whether it then works more efficiently than usual, so as to send a greater supply of blood to all parts of the body; for the skin instantly becomes pale, as during incipient faintness. This paleness of the surface, however, is probably in large part, or exclusively, due to the vasomotor centre being affected in such a manner as to cause the contraction of the small arteries of the skin. That the skin is much affected under the sense of great fear, we see in the marvellous and inexplicable manner in which perspiration immediately exudes from it. This exudation is all the more remarkable, as the surface is then cold, and hence the term a cold sweat; whereas, the sudorific glands are properly excited into action when the surface is heated. The hairs also on the skin stand erect; and the superficial muscles shiver. In connection with the disturbed action of the heart, the breathing is hurried. The salivary glands act imperfectly; the mouth becomes dry,[16] and is often opened and shut. I have also noticed that under slight fear there is a strong tendency to yawn. One of the best-marked symptoms is the trembling of all the muscles of the body; and this is often first seen in the lips. From this cause, and from the dryness of the mouth, the voice becomes husky or

---

[16] Mr. Bain ('The Emotions and the Will,' 1865, p. 54) explains in the following manner the origin of the custom "of subjecting criminals in India to the ordeal of the morsel of rice. The accused is made to take a mouthful of rice, and after a little time to throw it out. If the morsel is quite dry, the party is believed to be guilty,—his own evil conscience operating to paralyse the salivating organs."

indistinct, or may altogether fail. "Obstupui, steteruntque comæ, et vox faucibus hæsit."

Of vague fear there is a well-known and grand description in Job:—"In thoughts from the visions of the night, when deep sleep falleth on men, fear came upon me, and trembling, which made all my bones to shake. Then a spirit passed before my face; the hair of my flesh stood up. It stood still, but I could not discern the form thereof: an image was before my eyes, there was silence, and I heard a voice, saying, Shall mortal man be more just than God? Shall a man be more pure than his Maker?" (Job iv. 13.)

As fear increases into an agony of terror, we behold, as under all violent emotions, diversified results. The heart beats wildly, or may fail to act and faintness ensue; there is a death-like pallor; the breathing is laboured; the wings of the nostrils are wildly dilated; "there is a gasping and convulsive motion of the lips, a tremor on the hollow cheek, a gulping and catching of the throat;"[17] the uncovered and protruding eyeballs are fixed on the object of terror; or they may roll restlessly from side to side, *huc illuc volvens oculos totumque pererrat*.[18] The pupils are said to be enormously dilated. All the muscles of the body may become rigid, or may be thrown into convulsive movements. The hands are alternately clenched and opened, often with a twitching movement. The arms may be protruded, as if to avert some dreadful danger, or may be thrown wildly over the head. The Rev. Mr. Hagenauer has seen this latter action in a terrified Australian. In other cases there is

---

[17] Sir C. Bell, Transactions of Royal Phil. Soc. 1822, p. 308. 'Anatomy of Expression,' p. 88 and pp. 164--169.

[18] See Moreau on the rolling of the eyes, in the edit. of 1820 of Lavater, tome iv. p. 263. Also, Gratiolet, De la Phys. p. 17.

a sudden and uncontrollable tendency to headlong flight; and so strong is this, that the boldest soldiers may be seized with a sudden panic.

As fear rises to an extreme pitch, the dreadful scream of terror is heard. Great beads of sweat stand on the skin. All the muscles of the body are relaxed. Utter prostration soon follows, and the mental powers fail. The intestines are affected. The sphincter muscles cease to act, and no longer retain the contents of the body.

Dr. J. Crichton Browne has given me so striking an account of intense fear in an insane woman, aged thirty-five, that the description though painful ought not to be omitted. When a paroxysm seizes her, she screams out, "This is hell!" "There is a black woman!" "I can't get out!"—and other such exclamations. When thus screaming, her movements are those of alternate tension and tremor. For one instant she clenches her hands, holds her arms out before her in a stiff semi-flexed position; then suddenly bends her body forwards, sways rapidly to and fro, draws her fingers through her hair, clutches at her neck, and tries to tear off her clothes. The sterno-cleido-mastoid muscles (which serve to bend the head on the chest) stand out prominently, as if swollen, and the skin in front of them is much wrinkled. Her hair, which is cut short at the back of her head, and is smooth when she is calm, now stands on end; that in front being dishevelled by the movements of her hands. The countenance expresses great mental agony. The skin is flushed over the face and neck, down to the clavicles, and the veins of the forehead and neck stand out like thick cords. The lower lip drops, and is somewhat everted. The mouth is kept half open, with the lower jaw projecting. The cheeks are hollow and deeply furrowed in curved lines running from the wings of the nostrils to the corners of the mouth. The nostrils themselves are raised and extended. The eyes are widely opened, and beneath them the skin appears swollen; the pupils are large. The forehead is wrinkled transversely in many folds, and at the inner extremities of the eyebrows it is strongly furrowed in diverging lines, produced by the powerful and persistent contraction of the corrugators.

Mr. Bell has also described [19] an agony of terror and of despair, which he witnessed in a murderer, whilst carried to the place of execution in Turin. "On each side of the car the officiating priests were seated; and in the centre sat the criminal himself. It was impossible to witness the condition of this unhappy wretch without terror; and yet, as if impelled by some strange infatuation, it was equally impossible not to gaze upon an object so wild, so full of horror. He seemed about thirty-five years of age; of large and muscular form; his countenance marked by strong and savage features; half naked, pale as death, agonized with terror, every limb strained in anguish, his hands clenched convulsively, the sweat breaking out on his bent and contracted brow, he kissed incessantly the figure of our Saviour, painted on the flag which was suspended before him; but with an agony of wildness and despair, of which nothing ever exhibited on the stage can give the slightest conception."

I will add only one other case, illustrative of a man utterly prostrated by terror. An atrocious murderer of two persons was brought into a hospital, under the mistaken impression that he had poisoned himself; and Dr. W. Ogle carefully watched him the next morning, while he was being handcuffed and taken away by the police. His pallor was extreme, and his prostration so great that

[19] 'Observations on Italy,' 1825, p. 48, as quoted in 'The Anatomy of Expression,' p. 168.

he was hardly able to dress himself. His skin perspired; and his eyelids and head drooped so much that it was impossible to catch even a glimpse of his eyes. His lower jaw hung down. There was no contraction of any facial muscle, and Dr. Ogle is almost certain that the hair did not stand on end, for he observed it narrowly, as it had been dyed for the sake of concealment.

With respect to fear, as exhibited by the various races of man, my informants agree that the signs are the same as with Europeans. They are displayed in an exaggerated degree with the Hindoos and natives of Ceylon. Mr. Geach has seen Malays when terrified turn pale and shake; and Mr. Brough Smyth states that a native Australian "being on one occasion much frightened, showed a complexion as nearly approaching to what we call paleness, as can well be conceived in the case of a very black man." Mr. Dyson Lacy has seen extreme fear shown in an Australian, by a nervous twitching of the hands, feet, and lips; and by the perspiration standing on the skin. Many savages do not repress the signs of fear so much as Europeans; and they often tremble greatly. With the Kafir, Gaika says, in his rather quaint English, the shaking "of the body is much experienced, and the eyes are widely open." With savages, the sphincter muscles are often relaxed, just as may be observed in much frightened dogs, and as I have seen with monkeys when terrified by being caught.

The erection of the hair.—Some of the signs of fear deserve a little further consideration. Poets continually speak of the hair standing on end; Brutus says to the ghost of Cæsar, "that mak'st my blood cold, and my hair to stare." And Cardinal Beaufort, after the murder of Gloucester exclaims, "Comb down his hair; look, look, it stands upright." As I did not feel sure whether writers of fiction might not have applied to man what they had often observed in animals, I begged for information from Dr. Crichton Browne with respect to the insane. He states in answer that he has repeatedly seen their hair erected under the influence of sudden and extreme terror. For instance, it is occasionally necessary to inject morphia under the skin of an insane woman, who dreads the operation extremely, though it causes very little pain; for she believes that poison is being introduced into her system, and that her bones will be softened, and her flesh turned into dust. She becomes deadly pale; her limbs are stiffened by a sort of tetanic spasm, and her hair is partially erected on the front of the head.

Dr. Browne further remarks that the bristling of the hair which is so common in the insane, is not always associated with terror. It is perhaps most frequently seen in chronic maniacs, who rave incoherently and have destructive impulses; but it is during their paroxysms of violence that the bristling is most observable. The fact of the hair becoming erect under the influence both of rage and fear agrees perfectly with what we have seen in the lower animals. Dr. Browne adduces several cases in evidence. Thus with a man now in the Asylum, before the recurrence of each maniacal paroxysm, "the hair rises up from his forehead like the mane of a Shetland pony." He has sent me photographs of two women, taken in the intervals between their paroxysms, and he adds with respect to one of these women, "that the state of her hair is a sure and convenient criterion of her mental condition." I have had one of these photographs copied, and the engraving gives, if viewed from a little distance, a faithful representation of the original, with the exception that the hair appears rather too coarse and too much curled. The extraordinary condition of the hair in the insane is due, not only to its erection, but to its dryness and harshness, consequent on the subcutaneous glands failing to act. Dr. Bucknill has said [20] that a

FIG. 19.—From a photograph of an insane woman, to show the condition of her hair.

lunatic "is a lunatic to his finger's ends;" he might have added, and often to the extremity of each particular hair.

Dr. Browne mentions as an empirical confirmation of the relation which exists in the insane between the state of their hair and minds, that the wife of a medical man, who has charge of a lady suffering from acute melancholia, with a strong fear of death, for herself, her husband and children, reported verbally to him the day before receiving my letter as follows, "I think Mrs. —— will soon improve, for her hair is getting smooth; and I always notice that our patients get better whenever their hair ceases to be rough and unmanageable."

Dr. Browne attributes the persistently rough condi-

[20] Quoted by Dr. Maudsley, 'Body and Mind,' 1870, p. 41.

tion of the hair in many insane patients, in part to their minds being always somewhat disturbed, and in part to the effects of habit,—that is, to the hair being frequently and strongly erected during their many recurrent paroxysms. In patients in whom the bristling of the hair is extreme, the disease is generally permanent and mortal; but in others, in whom the bristling is moderate, as soon as they recover their health of mind the hair recovers its smoothness.

In a previous chapter we have seen that with animals the hairs are erected by the contraction of minute, unstriped, and involuntary muscles, which run to each separate follicle. In addition to this action, Mr. J. Wood has clearly ascertained by experiment, as he informs me, that with man the hairs on the front of the head which slope forwards, and those on the back which slope backwards, are raised in opposite directions by the contraction of the occipito-frontalis or scalp muscle. So that this muscle seems to aid in the erection of the hairs on the head of man, in the same manner as the homologous *panniculus carnosus* aids, or takes the greater part, in the erection of the spines on the backs of some of the lower animals.

Contraction of the platysma myoides muscle.—This muscle is spread over the sides of the neck, extending downwards to a little beneath the collar-bones, and upwards to the lower part of the cheeks. A portion, called the risorius, is represented in the woodcut (M) fig. 2. The contraction of this muscle draws the corners of the mouth and the lower parts of the cheeks downwards and backwards. It produces at the same time divergent, longitudinal, prominent ridges on the sides of the neck in the young; and, in old thin persons, fine transverse wrinkles. This muscle is sometimes said not to be under the control of the will; but almost every one, if told to draw the corners of his mouth backwards and downwards with great force, brings it into action. I have, however, heard of a man who can voluntarily act on it only on one side of his neck.

Sir C. Bell [21] and others have stated that this muscle is strongly contracted under the influence of fear; and Duchenne insists so strongly on its importance in the expression of this emotion, that he calls it the *muscle of fright.*[22] He admits, however, that its contraction is quite inexpressive unless associated with widely open eyes and mouth. He has given a photograph (copied and reduced in the accompanying woodcut) of the same old man as on former occasions, with his eyebrows strongly raised, his mouth opened, and the platysma contracted, all by means of galvanism. The original photograph was shown to twenty-four persons, and they were separately asked, without any explanation being given, what expression was intended: twenty instantly answered, "intense fright" or "horror;" three said pain, and one extreme discomfort. Dr. Duchenne has given another photograph of the same old man, with the platysma contracted, the eyes and mouth opened, and the eyebrows rendered oblique, by means of galvanism. The expression thus induced is very striking (see Plate VII. fig. 2); the obliquity of the eyebrows adding the appearance of great mental distress. The original was shown to fifteen persons; twelve answered terror or horror, and three agony or great suffering. From these cases, and from an examination of the other photographs given by Dr. Duchenne, together with his remarks thereon, I think there can be little doubt that the contraction of

---

[21] 'Anatomy of Expression,' p. 168.

[22] Mécanisme de la Phys. Humaine, Album, Légende xi.

the platysma does add greatly to the expression of fear. Nevertheless this muscle ought hardly to be called that of fright, for its contraction is certainly not a necessary concomitant of this state of mind.

FIG. 20.—Terror, from a photograph by Dr. Duchenne.

A man may exhibit extreme terror in the plainest manner by death-like pallor, by drops of perspiration on his skin, and by utter prostration, with all the muscles of his body, including the platysma, completely relaxed. Although Dr. Browne has often seen this muscle quivering and contracting in the insane, he has not been able to connect its action with any emotional condition in them, though he carefully attended to patients suffering from great fear. Mr. Nicol, on the other hand, has observed three cases in which this muscle appeared to be more or less permanently contracted under the influence of melancholia, associated with much dread; but in one of these cases, various other muscles about the neck and head were subject to spasmodic contractions.

Dr. W. Ogle observed for me in one of the London hospitals about twenty patients, just before they were put under the influence of chloroform for operations. They exhibited some trepidation, but no great terror. In only four of the cases was the platysma visibly contracted; and it did not begin to contract until the patients began to cry. The muscle seemed to contract at the moment of each deep-drawn inspiration; so that it is very doubtful whether the contraction depended at all on the emotion of fear. In a fifth case, the patient, who was not chloroformed, was much terrified; and his platysma was more forcibly and persistently contracted than in the other cases. But even here there is room for doubt, for the muscle which appeared to be unusually developed, was seen by Dr. Ogle to contract as the man moved his head from the pillow, after the operation was over.

As I felt much perplexed why, in any case, a superficial muscle on the neck should be especially affected by fear, I applied to my many obliging correspondents for information about the contraction of this muscle under other circumstances. It would be superfluous to give all the answers which I have received. They show that this muscle acts, often in a variable manner and degree, under many different conditions. It is violently contracted in hydrophobia, and in a somewhat less de-

PLATE VII

1

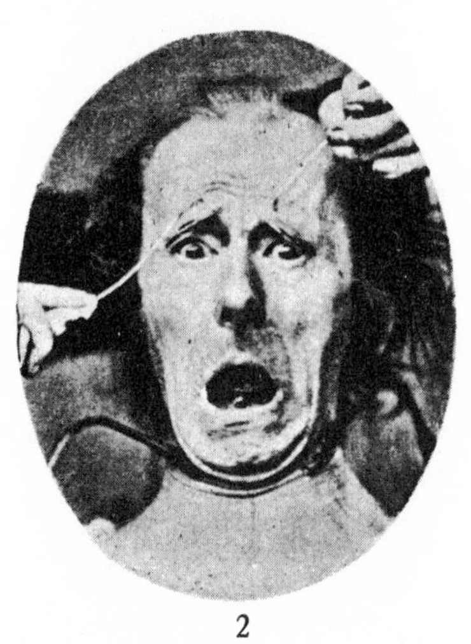

2

gree in lockjaw; sometimes in a marked manner during the insensibility from chloroform. Dr. W. Ogle observed two male patients, suffering from such difficulty in breathing, that the trachea had to be opened, and in both the platysma was strongly contracted. One of these men overheard the conversation of the surgeons surrounding him, and when he was able to speak, declared that he had not been frightened. In some other cases of extreme difficulty of respiration, though not requiring tracheotomy, observed by Drs. Ogle and Langstaff, the platysma was not contracted.

Mr. J. Wood, who has studied with such care the muscles of the human body, as shown by his various publications, has often seen the platysma contracted in vomiting, nausea, and disgust; also in children and adults under the influence of rage,—for instance, in Irishwomen, quarrelling and brawling together with angry gesticulations. This may possibly have been due to their high and angry tones; for I know a lady, an excellent musician, who, in singing certain high notes, always contracts her platysma. So does a young man, as I have observed, in sounding certain notes on the flute. Mr. J. Wood informs me that he has found the platysma best developed in persons with thick necks and broad shoulders; and that in families inheriting these peculiarities, its development is usually associated with much voluntary power over the homologous occipito-frontalis muscle, by which the scalp can be moved.

None of the foregoing cases appear to throw any light on the contraction of the platysma from fear; but it is different, I think, with the following cases. The gentleman before referred to, who can voluntarily act on this muscle only on one side of his neck, is positive that it contracts on both sides whenever he is startled. Evidence has already been given showing that this muscle sometimes contracts, perhaps for the sake of opening the mouth widely, when the breathing is rendered difficult by disease, and during the deep inspirations of crying-fits before an operation. Now, whenever a person starts at any sudden sight or sound, he instantaneously draws a deep breath; and thus the contraction of the platysma may possibly have become associated with the sense of fear. But there is, I believe, a more efficient relation. The first sensation of fear, or the imagination of something dreadful, commonly excites a shudder. I have caught myself giving a little involuntary shudder at a painful thought, and I distinctly perceived that my platysma contracted; so it does if I simulate a shudder. I have asked others to act in this manner; and in some the muscle contracted, but not in others. One of my sons, whilst getting out of bed, shuddered from the cold, and, as he happened to have his hand on his neck, he plainly felt that this muscle strongly contracted. He then voluntarily shuddered, as he had done on former occasions, but the platysma was not then affected. Mr. J. Wood has also several times observed this muscle contracting in patients, when stripped for examination, and who were not frightened, but shivered slightly from the cold. Unfortunately I have not been able to ascertain whether, when the whole body shakes, as in the cold stage of an ague fit, the platysma contracts. But as it certainly often contracts during a shudder; and as a shudder or shiver often accompanies the first sensation of fear, we have, I think, a clue to its action in this latter case.[23] Its contraction, however, is not an invariable

---

[23] Duchenne takes, in fact, this view (ibid. p. 45), as he attributes the contraction of the platysma to the shivering of fear (*frisson de la peur*); but he elsewhere compares the action with that which causes the hair of frightened quadrupeds to stand erect; and this can hardly be considered as quite correct.

concomitant of fear; for it probably never acts under the influence of extreme, prostrating terror.

Dilatation of the Pupils.—Gratiolet repeatedly insists [24] that the pupils are enormously dilated whenever terror is felt. I have no reason to doubt the accuracy of this statement, but have failed to obtain confirmatory evidence, excepting in the one instance before given of an insane woman suffering from great fear. When writers of fiction speak of the eyes being widely dilated, I presume that they refer to the eyelids. Munro's statement,[25] that with parrots the iris is affected by the passions, independently of the amount of light, seems to bear on this question; but Professor Donders informs me, that he has often seen movements in the pupils of these birds which he thinks may be related to their power of accommodation to distance, in nearly the same manner as our own pupils contract when our eyes converge for near vision. Gratiolet remarks that the dilated pupils appear as if they were gazing into profound darkness. No doubt the fears of man have often been excited in the dark; but hardly so often or so exclusively, as to account for a fixed and associated habit having thus arisen. It seems more probable, assuming that Gratiolet's statement is correct, that the brain is directly affected by the powerful emotion of fear and reacts on the pupils; but Professor Donders informs me that this is an extremely complicated subject. I may add, as possibly throwing light on the subject, that Dr. Fyffe, of Netley Hospital, has observed in two patients that the pupils were distinctly dilated during the cold stage of an ague fit. Professor Donders has also often seen dilatation of the pupils in incipient faintness.

---

[24] 'De la Physionomie,' pp. 51, 256, 346.
[25] As quoted in White's 'Gradation in Man,' p. 57.

Horror.—The state of mind expressed by this term implies terror, and is in some cases almost synonymous with it. Many a man must have felt, before the blessed discovery of chloroform, great horror at the thought of an impending surgical operation. He who dreads, as well as hates a man, will feel, as Milton uses the word, a horror of him. We feel horror if we see any one, for instance a child, exposed to some instant and crushing danger. Almost every one would experience the same feeling in the highest degree in witnessing a man being tortured or going to be tortured. In these cases there is no danger to ourselves; but from the power of the imagination and of sympathy we put ourselves in the position of the sufferer, and feel something akin to fear.

Sir C. Bell remarks,[26] that "horror is full of energy; the body is in the utmost tension, not unnerved by fear." It is, therefore, probable that horror would generally be accompanied by the strong contraction of the brows; but as fear is one of the elements, the eyes and mouth would be opened, and the eyebrows would be raised, as far as the antagonistic action of the corrugators permitted this movement. Duchenne has given a photograph [27] (fig. 21) of the same old man as before, with his eyes somewhat staring, the eyebrows partially raised, and at the same time strongly contracted, the mouth opened, and the platysma in action, all effected by the means of galvanism. He considers that the expression thus produced shows extreme terror with horrible pain or torture. A tortured man, as long as his sufferings allowed him to feel any dread for the future, would probably exhibit horror in an extreme degree. I have shown the original of this photograph to twenty-three persons of both sexes

---

[26] 'Anatomy of Expression,' p. 169.

[27] 'Mécanisme de la Physionomie,' Album, pl. 65, pp. 44, 45.

and various ages; and thirteen immediately answered horror, great pain, torture, or agony; three answered extreme fright; so that sixteen answered nearly in accordance with Duchenne's belief. Six, however, said anger, guided no doubt, by the strongly contracted brows, and overlooking the peculiarly opened mouth. One said disgust. On the whole, the evidence indicates that we have here a fairly good representation of horror and agony. The photograph before referred to (Pl. VII. fig. 2) likewise exhibits horror; but in this the oblique eyebrows indicate great mental distress in place of energy.

Horror is generally accompanied by various gestures, which differ in different individuals. Judging from pictures, the whole body is often turned away or shrinks; or the arms are violently protruded as if to push away some dreadful object. The most frequent gesture, as far as can be inferred from the action of persons who endeavour to express a vividly-imagined scene of horror, is the raising of both shoulders, with the bent arms pressed closely against the sides or chest. These movements are nearly the same with those commonly made when we feel very cold; and they are generally accompanied by a shudder, as well as by a deep expiration or inspiration, according as the chest happens at the time to be expanded or contracted. The sounds thus made are expressed by words like *uh* or *ugh*.[28] It is not, however, obvious why, when we feel cold or express a sense of horror, we press our bent arms against our bodies, raise our shoulders, and shudder.

---

[28] See remarks to this effect by Mr. Wedgwood, in the Introduction to his 'Dictionary of English Etymology,' 2nd edit. 1872, p. xxxvii. He shows by intermediate forms that the sounds here referred to have probably given rise to many words, such as *ugly*, *huge*, &c.

FIG. 21.—Horror and Agony, copied from a photograph by Dr. Duchenne.

Conclusion.—I have now endeavoured to describe the diversified expressions of fear, in its gradations from mere attention to a start of surprise, into extreme terror and horror. Some of the signs may be accounted for through the principles of habit, association, and inheritance,—such as the wide opening of the mouth and eyes, with upraised eyebrows, so as to see as quickly as possible all around us, and to hear distinctly whatever sound may reach our ears. For we have thus habitually prepared ourselves to discover and encounter any danger. Some of the other signs of fear may likewise be accounted for, at least in part, through these same principles. Men, during numberless generations, have endeavoured to escape from their enemies or danger by headlong flight, or by violently struggling with them; and such great exertions will have caused the heart to beat rapidly, the breathing to be hurried, the chest to heave, and the nostrils to be dilated. As these exertions have often been prolonged to the last extremity, the final result will have been utter prostration, pallor, perspiration, trembling of all the muscles, or their complete relaxation. And now, whenever the emotion of fear is strongly felt, though it may not lead to any exertion, the same results tend to reappear, through the force of inheritance and association.

Nevertheless, it is probable that many or most of the above symptoms of terror, such as the beating of the heart, the trembling of the muscles, cold perspiration, &c., are in large part directly due to the disturbed or interrupted transmission of nerve-force from the cerebrospinal system to various parts of the body, owing to the mind being so powerfully affected. We may confidently look to this cause, independently of habit and association, in such cases as the modified secretions of the intestinal canal, and the failure of certain glands to act. With respect to the involuntary bristling of the hair, we have good reason to believe that in the case of animals this action, however it may have originated, serves, together with certain voluntary movements, to make them appear terrible to their enemies; and as the same involuntary and voluntary actions are performed by animals nearly related to man, we are led to believe that man has retained through inheritance a relic of them, now become useless. It is certainly a remarkable fact, that the minute unstriped muscles, by which the hairs thinly scattered over man's almost naked body are erected, should have been preserved to the present day; and that they should still contract under the same emotions, namely, terror and rage, which cause the hairs to stand on end in the lower members of the Order to which man belongs.

CHAPTER XIII.

SELF-ATTENTION—SHAME—SHYNESS—MODESTY: BLUSHING.

Nature of a blush—Inheritance—The parts of the body most affected—Blushing in the various races of man—Accompanying gestures—Confusion of mind—Causes of blushing—Self-attention, the fundamental element—Shyness—Shame, from broken moral laws and conventional rules—Modesty—Theory of blushing—Recapitulation.

BLUSHING is the most peculiar and the most human of all expressions. Monkeys redden from passion, but it would require an overwhelming amount of evidence to make us believe that any animal could blush. The reddening of the face from a blush is due to the relaxation of the muscular coats of the small arteries, by which the capillaries become filled with blood; and this depends on the proper vaso-motor centre being affected. No doubt if there be at the same time much mental agitation, the general circulation will be affected; but it is not due to the action of the heart that the network of minute vessels covering the face becomes under a sense of shame gorged with blood. We can cause laughing by tickling the skin, weeping or frowning by a blow, trembling from the fear of pain, and so forth; but we cannot cause a blush, as Dr. Burgess remarks,[1] by

[1] 'The Physiology or Mechanism of Blushing,' 1839, p. 156. I shall have occasion often to quote this work in the present chapter.

any physical means,—that is by any action on the body. It is the mind which must be affected. Blushing is not only involuntary; but the wish to restrain it, by leading to self-attention actually increases the tendency.

The young blush much more freely than the old, but not during infancy,[2] which is remarkable, as we know that infants at a very early age redden from passion. I have received authentic accounts of two little girls blushing at the ages of between two and three years; and of another sensitive child, a year older, blushing, when reproved for a fault. Many children, at a somewhat more advanced age blush in a strongly marked manner. It appears that the mental powers of infants are not as yet sufficiently developed to allow of their blushing. Hence, also, it is that idiots rarely blush. Dr. Crichton Browne observed for me those under his care, but never saw a genuine blush, though he has seen their faces flush, apparently from joy, when food was placed before them, and from anger. Nevertheless some, if not utterly degraded, are capable of blushing. A microcephalous idiot, for instance, thirteen years old, whose eyes brightened a little when he was pleased or amused, has been described by Dr. Behn,[3] as blushing and turning to one side, when undressed for medical examination.

Women blush much more than men. It is rare to see an old man, but not nearly so rare to see an old woman blushing. The blind do not escape. Laura Bridgman, born in this condition, as well as completely

---

[2] Dr. Burgess, ibid. p. 56. At p. 33 he also remarks on women blushing more freely than men, as stated below.

[3] Quoted by Vogt, 'Mémoire sur les Microcéphales,' 1867, p. 20. Dr. Burgess (ibid. p. 56) doubts whether idiots ever blush.

deaf, blushes.[4] The Rev. R. H. Blair, Principal of the Worcester College, informs me that three children born blind, out of seven or eight then in the Asylum, are great blushers. The blind are not at first conscious that they are observed, and it is a most important part of their education, as Mr. Blair informs me, to impress this knowledge on their minds; and the impression thus gained would greatly strengthen the tendency to blush, by increasing the habit of self-attention.

The tendency to blush is inherited. Dr. Burgess gives the case [5] of a family consisting of a father, mother, and ten children, all of whom, without exception, were prone to blush to a most painful degree. The children were grown up; "and some of them were sent to travel in order to wear away this diseased sensibility, but nothing was of the slightest avail." Even peculiarities in blushing seem to be inherited. Sir James Paget, whilst examining the spine of a girl, was struck at her singular manner of blushing; a big splash of red appeared first on one cheek, and then other splashes, variously scattered over the face and neck. He subsequently asked the mother whether her daughter always blushed in this peculiar manner; and was answered, "Yes, she takes after me." Sir J. Paget then perceived that by asking this question he had caused the mother to blush; and she exhibited the same peculiarity as her daughter.

In most cases the face, ears and neck are the sole parts which redden; but many persons, whilst blushing intensely, feel that their whole bodies grow hot and tingle; and this shows that the entire surface must be in some manner affected. Blushes are said sometimes

[4] Lieber 'On the Vocal Sounds,' &c.; Smithsonian Contributions, 1851, vol. ii. p. 6.

[5] Ibid. p. 182.

to commence on the forehead, but more commonly on the cheeks, afterwards spreading to the ears and neck.[6] In two Albinos examined by Dr. Burgess, the blushes commenced by a small circumscribed spot on the cheeks, over the parotidean plexus of nerves, and then increased into a circle; between this blushing circle and the blush on the neck there was an evident line of demarcation; although both arose simultaneously. The retina, which is naturally red in the Albino, invariably increased at the same time in redness.[7] Every one must have noticed how easily after one blush fresh blushes chase each other over the face. Blushing is preceded by a peculiar sensation in the skin. According to Dr. Burgess the reddening of the skin is generally succeeded by a slight pallor, which shows that the capillary vessels contract after dilating. In some rare cases paleness instead of redness is caused under conditions which would naturally induce a blush. For instance, a young lady told me that in a large and crowded party she caught her hair so firmly on the button of a passing servant, that it took some time before she could be extricated; from her sensations she imagined that she had blushed crimson; but was assured by a friend that she had turned extremely pale.

I was desirous to learn how far down the body blushes extend; and Sir J. Paget, who necessarily has frequent opportunities for observation, has kindly attended to this point for me during two or three years. He finds that with women who blush intensely on the face, ears, and nape of neck, the blush does not commonly extend any lower down the body. It is rare to see it as low down as the collar-bones and shoulder-blades; and he has never himself seen a single instance in which it

---

[6] Moreau, in edit. of 1820 of Lavater, vol. iv. p. 303.
[7] Burgess, ibid. p. 38, on paleness after blushing, p. 177.

extended below the upper part of the chest. He has also noticed that blushes sometimes die away downwards, not gradually and insensibly, but by irregular ruddy blotches. Dr. Langstaff has likewise observed for me several women whose bodies did not in the least redden while their faces were crimsoned with blushes. With the insane, some of whom appear to be particularly liable to blushing, Dr. J. Crichton Browne has several times seen the blush extend as far down as the collar-bones, and in two instances to the breasts. He gives me the case of a married woman, aged twenty-seven, who suffered from epilepsy. On the morning after her arrival in the Asylum, Dr. Browne, together with his assistants, visited her whilst she was in bed. The moment that he approached, she blushed deeply over her cheeks and temples; and the blush spread quickly to her ears. She was much agitated and tremulous. He unfastened the collar of her chemise in order to examine the state of her lungs; and then a brilliant blush rushed over her chest, in an arched line over the upper third of each breast, and extended downwards between the breasts nearly to the ensiform cartilage of the sternum. This case is interesting, as the blush did not thus extend downwards until it became intense by her attention being drawn to this part of her person. As the examination proceeded she became composed, and the blush disappeared; but on several subsequent occasions the same phenomena were observed.

The foregoing facts show that, as a general rule, with English women, blushing does not extend beneath the neck and upper part of the chest. Nevertheless Sir J. Paget informs me that he has lately heard of a case, on which he can fully rely, in which a little girl, shocked by what she imagined to be an act of indelicacy, blushed all over her abdomen and the upper parts of her legs.

Moreau also [8] relates, on the authority of a celebrated painter, that the chest, shoulders, arms, and whole body of a girl, who unwillingly consented to serve as a model, reddened when she was first divested of her clothes.

It is a rather curious question why, in most cases the face, ears, and neck alone redden, inasmuch as the whole surface of the body often tingles and grows hot. This seems to depend, chiefly, on the face and adjoining parts of the skin having been habitually exposed to the air, light, and alternations of temperature, by which the small arteries not only have acquired the habit of readily dilating and contracting, but appear to have become unusually developed in comparison with other parts of the surface.[9] It is probably owing to this same cause, as M. Moreau and Dr. Burgess have remarked, that the face is so liable to redden under various circumstances, such as a fever-fit, ordinary heat, violent exertion, anger, a slight blow, &c.; and on the other hand that it is liable to grow pale from cold and fear, and to be discoloured during pregnancy. The face is also particularly liable to be affected by cutaneous complaints, by small-pox, erysipelas, &c. This view is likewise supported by the fact that the men of certain races, who habitually go nearly naked, often blush over their arms and chests and even down to their waists. A lady, who is a great blusher, informs Dr. Crichton Browne, that when she feels ashamed or is agitated, she blushes over her face, neck, wrists, and hands,—that is, over all the exposed portions of her skin. Nevertheless it may be doubted whether the habitual exposure of the skin of the face and neck, and its consequent power of reaction under stimulants of all kinds, is by itself sufficient to account for the much

---

[8] See Lavater, edit. of 1820, vol. iv. p. 303.

[9] Burgess, ibid. pp. 114, 122. Moreau in Lavater, ibid. vol. iv. p. 293.

greater tendency in English women of these parts than of others to blush; for the hands are well supplied with nerves and small vessels, and have been as much exposed to the air as the face or neck, and yet the hands rarely blush. We shall presently see that the attention of the mind having been directed much more frequently and earnestly to the face than to any other part of the body, probably affords a sufficient explanation.

Blushing in the various races of man.—The small vessels of the face become filled with blood, from the emotion of shame, in almost all the races of man, though in the very dark races no distinct change of colour can be perceived. Blushing is evident in all the Aryan nations of Europe, and to a certain extent with those of India. But Mr. Erskine has never noticed that the necks of the Hindoos are decidedly affected. With the Lepchas of Sikhim, Mr. Scott has often observed a faint blush on the cheeks, base of the ears, and sides of the neck, accompanied by sunken eyes and lowered head. This has occurred when he has detected them in a falsehood, or has accused them of ingratitude. The pale, sallow complexions of these men render a blush much more conspicuous than in most of the other natives of India. With the latter, shame, or it may be in part fear, is expressed, according to Mr. Scott, much more plainly by the head being averted or bent down, with the eyes wavering or turned askant, than by any change of colour in the skin.

The Semitic races blush freely, as might have been expected, from their general similitude to the Aryans. Thus with the Jews, it is said in the Book of Jeremiah (chap. vi. 15), "Nay, they were not at all ashamed, neither could they blush." Mrs. Asa Gray saw an Arab managing his boat clumsily on the Nile, and when laughed at by his companions, "he blushed quite to the back of his neck." Lady Duff Gordon remarks that a young Arab blushed on coming into her presence.[10]

Mr. Swinhoe has seen the Chinese blushing, but he thinks it is rare; yet they have the expression "to redden with shame." Mr. Geach informs me that the Chinese settled in Malacca and the native Malays of the interior both blush. Some of these people go nearly naked, and he particularly attended to the downward extension of the blush. Omitting the cases in which the face alone was seen to blush, Mr. Geach observed that the face, arms, and breast of a Chinaman, aged 24 years, reddened from shame; and with another Chinese, when asked why he had not done his work in better style, the whole body was similarly affected. In two Malays [11] he saw the face, neck, breast, and arms blushing; and in a third Malay (a Bugis) the blush extended down to the waist.

The Polynesians blush freely. The Rev. Mr. Stack has seen hundreds of instances with the New Zealanders. The following case is worth giving, as it relates to an old man who was unusually dark-coloured and partly tattooed. After having let his land to an Englishman for a small yearly rental, a strong passion seized him to buy a gig, which had lately become the fashion with the Maoris. He consequently wished to draw all the rent for four years from his tenant, and consulted Mr. Stack whether he could do so. The man was old, clumsy, poor, and ragged, and the idea of his driving himself about in his carriage for display amused Mr. Stack so much that he could not help bursting out into a laugh; and then "the old man blushed up to the roots of his hair."

[10] 'Letters from Egypt,' 1865, p. 66. Lady Gordon is mistaken when she says Malays and Mulattoes never blush.

[11] Capt. Osborn ('Quedah,' p. 199), in speaking of a Malay, whom he reproached for cruelty, says he was glad to see that the man blushed.

Forster says that "you may easily distinguish a spreading blush" on the cheeks of the fairest women in Tahiti.[12] The natives also of several of the other archipelagoes in the Pacific have been seen to blush.

Mr. Washington Matthews has often seen a blush on the faces of the young squaws belonging to various wild Indian tribes of North America. At the opposite extremity of the continent in Tierra del Fuego, the natives, according to Mr. Bridges, "blush much, but chiefly in regard to women; but they certainly blush also at their own personal appearance." This latter statement agrees with what I remember of the Fuegian, Jemmy Button, who blushed when he was quizzed about the care which he took in polishing his shoes, and in otherwise adorning himself. With respect to the Aymara Indians on the lofty plateaus of Bolivia, Mr. Forbes says,[13] that from the colour of their skins it is impossible that their blushes should be as clearly visible as in the white races; still under such circumstances as would raise a blush in us, "there can always be seen the same expression of modesty or confusion; and even in the dark, a rise of temperature of the skin of the face can be felt, exactly as occurs in the European." With the Indians who in-

---

[12] J. R. Forster, 'Observations during a Voyage round the World,' 4to, 1778, p. 229. Waitz gives ('Introduction to Anthropology,' Eng. translat. 1863, vol. i. p. 135) references for other islands in the Pacific. See, also, Dampier 'On the Blushing of the Tunquinese' (vol. ii. p. 40); but I have not consulted this work. Waitz quotes Bergmann, that the Kalmucks do not blush, but this may be doubted after what we have seen with respect to the Chinese. He also quotes Roth, who denies that the Abyssinians are capable of blushing. Unfortunately, Capt. Speedy, who lived so long with the Abyssinians, has not answered my inquiry on this head. Lastly, I must add that the Rajah Brooke has never observed the least sign of a blush with the Dyaks of Borneo; on the contrary under circumstances which would excite a blush in us, they assert "that they feel the blood drawn from their faces."

[13] Transact. of the Ethnological Soc. 1870, vol. ii. p. 16.

habit the hot, equable, and damp parts of South America, the skin apparently does not answer to mental excitement so readily as with the natives of the northern and southern parts of the continent, who have long been exposed to great vicissitudes of climate; for Humboldt quotes without a protest the sneer of the Spaniard, "How can those be trusted, who know not how to blush?"[14] Von Spix and Martius, in speaking of the aborigines of Brazil, assert that they cannot properly be said to blush; "it was only after long intercourse with the whites, and after receiving some education, that we perceived in the Indians a change of colour expressive of the emotions of their minds."[15] It is, however, incredible that the power of blushing could have thus originated; but the habit of self-attention, consequent on their education and new course of life, would have much increased any innate tendency to blush.

Several trustworthy observers have assured me that they have seen on the faces of negroes an appearance resembling a blush, under circumstances which would have excited one in us, though their skins were of an ebony-black tint. Some describe it as blushing brown, but most say that the blackness becomes more intense. An increased supply of blood in the skin seems in some manner to increase its blackness; thus certain exanthematous diseases cause the affected places in the negro to appear blacker, instead of, as with us, redder.[16] The skin, perhaps, from being rendered more tense by the

---

[14] Humboldt, 'Personal Narrative,' Eng. translat. vol. iii. p. 229.

[15] Quoted by Prichard, Phys. Hist. of Mankind, 4th edit. 1851, vol. i. p. 271.

[16] See, on this head, Burgess, ibid. p. 32. Also Waitz, 'Introduction to Anthropology,' Eng. edit. vol. i. p. 135. Moreau gives a detailed account ('Lavater,' 1820, tom. iv. p. 302) of the blushing of a Madagascar negress-slave when forced by her brutal master to exhibit her naked bosom.

filling of the capillaries, would reflect a somewhat different tint to what it did before. That the capillaries of the face in the negro become filled with blood, under the emotion of shame, we may feel confident; because a perfectly characterized albino negress, described by Buffon,[17] showed a faint tinge of crimson on her cheeks when she exhibited herself naked. Cicatrices of the skin remain for a long time white in the negro, and Dr. Burgess, who had frequent opportunities of observing a scar of this kind on the face of a negress, distinctly saw that it "invariably became red whenever she was abruptly spoken to, or charged with any trivial offence."[18] The blush could be seen proceeding from the circumference of the scar towards the middle, but it did not reach the centre. Mulattoes are often great blushers, blush succeeding blush over their faces. From these facts there can be no doubt that negroes blush, although no redness is visible on the skin.

I am assured by Gaika and by Mrs. Barber that the Kafirs of South Africa never blush; but this may only mean that no change of colour is distinguishable. Gaika adds that under the circumstances which would make a European blush, his countrymen "look ashamed to keep their heads up."

It is asserted by four of my informants that the Australians, who are almost as black as negroes, never blush. A fifth answers doubtfully, remarking that only a very strong blush could be seen, on account of the dirty state of their skins. Three observers state that they do blush;[19] Mr. S. Wilson adding that this is noticeable

---

[17] Quoted by Prichard, Phys. Hist. of Mankind, 4th edit. 1851, vol. i. p. 225.

[18] Burgess, ibid. p. 31. On mulattoes blushing, see p. 33. I have received similar accounts with respect to mulattoes.

[19] Barrington also says that the Australians of New South Wales blush, as quoted by Waitz, ibid. p. 135.

only under a strong emotion, and when the skin is not too dark from long exposure and want of cleanliness. Mr. Lang answers, "I have noticed that shame almost always excites a blush, which frequently extends as low as the neck." Shame is also shown, as he adds, "by the eyes being turned from side to side." As Mr. Lang was a teacher in a native school, it is probable that he chiefly observed children; and we know that they blush more than adults. Mr. G. Taplin has seen half-castes blushing, and he says that the aborigines have a word expressive of shame. Mr. Hagenauer, who is one of those who has never observed the Australians to blush, says that he has "seen them looking down to the ground on account of shame;" and the missionary, Mr. Bulmer, remarks that though "I have not been able to detect anything like shame in the adult aborigines, I have noticed that the eyes of the children, when ashamed, present a restless, watery appearance, as if they did not know where to look."

The facts now given are sufficient to show that blushing, whether or not there is any change of colour, is common to most, probably to all, of the races of man.

Movements and gestures which accompany Blushing.—Under a keen sense of shame there is a strong desire for concealment.[20] We turn away the whole body, more especially the face, which we endeavour in some manner to hide. An ashamed person can hardly endure to meet

[20] Mr. Wedgwood says (Dict. of English Etymology, vol. iii. 1865, p. 155) that the word shame "may well originate in the idea of shade or concealment, and may be illustrated by the Low German *scheme*, shade or shadow." Gratiolet (De la Phys. pp. 357--362) has a good discussion on the gestures accompanying shame; but some of his remarks seem to me rather fanciful. See, also, Burgess (ibid. pp. 69, 134) on the same subject.

the gaze of those present, so that he almost invariably casts down his eyes or looks askant. As there generally exists at the same time a strong wish to avoid the appearance of shame, a vain attempt is made to look direct at the person who causes this feeling; and the antagonism between these opposite tendencies leads to various restless movements in the eyes. I have noticed two ladies who, whilst blushing, to which they are very liable, have thus acquired, as it appears, the oddest trick of incessantly blinking their eyelids with extraordinary rapidity. An intense blush is sometimes accompanied by a slight effusion of tears; [21] and this, I presume, is due to the lacrymal glands partaking of the increased supply of blood, which we know rushes into the capillaries of the adjoining parts, including the retina.

Many writers, ancient and modern, have noticed the foregoing movements; and it has already been shown that the aborigines in various parts of the world often exhibit their shame by looking downwards or askant, or by restless movements of their eyes. Ezra cries out (ch. ix. 6), "O, my God! I am ashamed, and blush to lift up my head to thee, my God." In Isaiah (ch. l. 6) we meet with the words, "I hid not my face from shame." Seneca remarks (Epist. xi. 5) "that the Roman players hang down their heads, fix their eyes on the ground and keep them lowered, but are unable to blush in acting shame." According to Macrobius, who lived in the fifth century ('Saturnalia,' B. vii. c. 11), "Natural philosophers assert that nature being moved by shame spreads the blood before herself as a veil, as we

[21] Burgess, ibid. pp. 181, 182. Boerhaave also noticed (as quoted by Gratiolet, ibid. p. 361) the tendency to the secretion of tears during intense blushing. Mr. Bulmer, as we have seen, speaks of the "watery eyes" of the children of the Australian aborigines when ashamed.

see any one blushing often puts his hands before his face." Shakspeare makes Marcus ('Titus Andronicus,' act ii, sc. 5) say to his niece, "Ah! now thou turn'st away thy face for shame." A lady informs me that she found in the Lock Hospital a girl whom she had formerly known, and who had become a wretched castaway, and the poor creature, when approached, hid her face under the bed-clothes, and could not be persuaded to uncover it. We often see little children, when shy or ashamed, turn away, and still standing up, bury their faces in their mother's gown; or they throw themselves face downwards on her lap.

Confusion of mind.—Most persons, whilst blushing intensely, have their mental powers confused. This is recognized in such common expressions as "she was covered with confusion." Persons in this condition lose their presence of mind, and utter singularly inappropriate remarks. They are often much distressed, stammer, and make awkward movements or strange grimaces. In certain cases involuntary twitchings of some of the facial muscles may be observed. I have been informed by a young lady, who blushes excessively, that at such times she does not even know what she is saying. When it was suggested to her that this might be due to her distress from the consciousness that her blushing was noticed, she answered that this could not be the case, "as she had sometimes felt quite as stupid when blushing at a thought in her own room."

I will give an instance of the extreme disturbance of mind to which some sensitive men are liable. A gentleman, on whom I can rely, assured me that he had been an eye-witness of the following scene:—A small dinner-party was given in honour of an extremely shy man, who, when he rose to return thanks, rehearsed the speech, which he had evidently learnt by heart, in absolute silence, and did not utter a single word; but he acted as if he were speaking with much emphasis. His friends, perceiving how the case stood, loudly applauded the imaginary bursts of eloquence, whenever his gestures indicated a pause, and the man never discovered that he had remained the whole time completely silent. On the contrary, he afterwards remarked to my friend, with much satisfaction, that he thought he had succeeded uncommonly well.

When a person is much ashamed or very shy, and blushes intensely, his heart beats rapidly and his breathing is disturbed. This can hardly fail to affect the circulation of the blood within the brain, and perhaps the mental powers. It seems however doubtful, judging from the still more powerful influence of anger and fear on the circulation, whether we can thus satisfactorily account for the confused state of mind in persons whilst blushing intensely.

The true explanation apparently lies in the intimate sympathy which exists between the capillary circulation of the surface of the head and face, and that of the brain. On applying to Dr. J. Crichton Browne for information, he has given me various facts bearing on this subject. When the sympathetic nerve is divided on one side of the head, the capillaries on this side are relaxed and become filled with blood, causing the skin to redden and to grow hot, and at the same time the temperature within the cranium on the same side rises. Inflammation of the membranes of the brain leads to the engorgement of the face, ears, and eyes with blood. The first stage of an epileptic fit appears to be the contraction of the vessels of the brain, and the first outward manifestation is an extreme pallor of countenance. Erysipelas of the head commonly induces delirium. Even the relief given to a severe headache by burning the skin with strong lotion, depends, I presume, on the same principle.

Dr. Browne has often administered to his patients the vapour of the nitrite of amyl,[22] which has the singular property of causing vivid redness of the face in from thirty to sixty seconds. This flushing resembles blushing in almost every detail: it begins at several distinct points on the face, and spreads till it involves the whole surface of the head, neck, and front of the chest; but has been observed to extend only in one case to the abdomen. The arteries in the retina become enlarged; the eyes glisten, and in one instance there was a slight effusion of tears. The patients are at first pleasantly stimulated, but, as the flushing increases, they become confused and bewildered. One woman to whom the vapour had often been administered asserted that, as soon as she grew hot, she grew *muddled.* With persons just commencing to blush it appears, judging from their bright eyes and lively behaviour, that their mental powers are somewhat stimulated. It is only when the blushing is excessive that the mind grows confused. Therefore it would seem that the capillaries of the face are affected, both during the inhalation of the nitrite of amyl and during blushing, before that part of the brain is affected on which the mental powers depend.

Conversely when the brain is primarily affected, the circulation of the skin is so in a secondary manner. Dr. Browne has frequently observed, as he informs me, scattered red blotches and mottlings on the chests of epileptic patients. In these cases, when the skin on the thorax or abdomen is gently rubbed with a pencil or other object, or, in strongly-marked cases, is merely touched by the

[22] See also Dr. J. Crichton Browne's Memoir on this subject in the 'West Riding Lunatic Asylum Medical Report,' 1871, pp. 95--98.

finger, the surface becomes suffused in less than half a minute with bright red marks, which spread to some distance on each side of the touched point, and persist for several minutes. These are the *cerebral maculæ* of Trousseau; and they indicate, as Dr. Browne remarks, a highly modified condition of the cutaneous vascular system. If, then, there exists, as cannot be doubted, an intimate sympathy between the capillary circulation in that part of the brain on which our mental powers depend, and in the skin of the face, it is not surprising that the moral causes which induce intense blushing should likewise induce, independently of their own disturbing influence, much confusion of mind.

The Nature of the Mental States which induce Blushing.—These consist of shyness, shame, and modesty; the essential element in all being self-attention. Many reasons can be assigned for believing that originally self-attention directed to personal appearance, in relation to the opinion of others, was the exciting cause; the same effect being subsequently produced, through the force of association, by self-attention in relation to moral conduct. It is not the simple act of reflecting on our own appearance, but the thinking what others think of us, which excites a blush. In absolute solitude the most sensitive person would be quite indifferent about his appearance. We feel blame or disapprobation more acutely than approbation; and consequently depreciatory remarks or ridicule, whether of our appearance or conduct, causes us to blush much more readily than does praise. But undoubtedly praise and admiration are highly efficient: a pretty girl blushes when a man gazes intently at her, though she may know perfectly well that he is not depreciating her. Many children, as well as old and sensitive persons blush, when they are much praised.

Hereafter the question will be discussed, how it has arisen that the consciousness that others are attending to our personal appearance should have led to the capillaries, especially those of the face, instantly becoming filled with blood.

My reasons for believing that attention directed to personal appearance, and not to moral conduct, has been the fundamental element in the acquirement of the habit of blushing, will now be given. They are separately light, but combined possess, as it appears to me, considerable weight. It is notorious that nothing makes a shy person blush so much as any remark, however slight, on his personal appearance. One cannot notice even the dress of a woman much given to blushing, wihout causing her face to crimson. It is sufficient to stare hard at some persons to make them, as Coleridge remarks, blush,—"account for that he who can."[23]

With the two albinos observed by Dr. Burgess,[24] "the slightest attempt to examine their peculiarities invariably" caused them to blush deeply. Women are much more sensitive about their personal appearance than men are, especially elderly women in comparison with elderly men, and they blush much more freely. The young of both sexes are much more sensitive on this same head than the old, and they also blush much more freely than the old. Children at a very early age do not blush; nor do they show those other signs of self-consciousness which generally accompany blushing; and it is one of their chief charms that they think nothing about what others think of them. At this early age they will stare at a stranger with a fixed gaze and un-

---

[23] In a discussion on so-called animal magnetism in 'Table Talk,' vol. i.

[24] Ibid. p. 40.

blinking eyes, as on an inanimate object, in a manner which we elders cannot imitate.

It is plain to every one that young men and women are highly sensitive to the opinion of each other with reference to their personal appearance; and they blush incomparably more in the presence of the opposite sex than in that of their own.[25] A young man, not very liable to blush, will blush intensely at any slight ridicule of his appearance from a girl whose judgment on any important subject he would disregard. No happy pair of young lovers, valuing each other's admiration and love more than anything else in the world, probably ever courted each other without many a blush. Even the barbarians of Tierra del Fuego, according to Mr. Bridges, blush "chiefly in regard to women, but certainly also at their own personal appearance."

Of all parts of the body, the face is most considered and regarded, as is natural from its being the chief seat of expression and the source of the voice. It is also the chief seat of beauty and of ugliness, and throughout the world is the most ornamented.[26] The face, therefore, will have been subjected during many generations to much closer and more earnest self-attention than any other part of the body; and in accordance with the principle here advanced we can understand why it should be the most liable to blush. Although exposure to alternations of temperature, &c., has probably much increased the power of dilatation and contraction in the capillaries of the face and adjoining parts, yet this by

---

[25] Mr. Bain ('The Emotions and the Will,' 1865, p. 65) remarks on "the shyness of manners which is induced between the sexes . . . . from the influence of mutual regard, by the apprehension on either side of not standing well with the other."

[26] See, for evidence on this subject, 'The Descent of Man,' &c., vol. ii. pp. 71, 341.

itself will hardly account for these parts blushing much more than the rest of the body; for it does not explain the fact of the hands rarely blushing. With Europeans the whole body tingles slightly when the face blushes intensely; and with the races of men who habitually go nearly naked, the blushes extend over a much larger surface than with us. These facts are, to a certain extent, intelligible, as the self-attention of primeval man, as well as of the existing races which still go naked, will not have been so exclusively confined to their faces, as is the case with the people who now go clothed.

We have seen that in all parts of the world persons who feel shame for some moral delinquency, are apt to avert, bend down, or hide their faces, independently of any thought about their personal appearance. The object can hardly be to conceal their blushes, for the face is thus averted or hidden under circumstances which exclude any desire to conceal shame, as when guilt is fully confessed and repented of. It is, however, probable that primeval man before he had acquired much moral sensitiveness would have been highly sensitive about his personal appearance, at least in reference to the other sex, and he would consequently have felt distress at any depreciatory remarks about his appearance; and this is one form of shame. And as the face is the part of the body which is most regarded, it is intelligible that any one ashamed of his personal appearance would desire to conceal this part of his body. The habit having been thus acquired, would naturally be carried on when shame from strictly moral causes was felt; and it is not easy otherwise to see why under these circumstances there should be a desire to hide the face more than any other part of the body.

The habit, so general with every one who feels ashamed, of turning away, or lowering his eyes, or restlessly moving them from side to side, probably follows from each glance directed towards those present, bringing home the conviction that he is intently regarded; and he endeavours, by not looking at those present, and especially not at their eyes, momentarily to escape from this painful conviction.

Shyness.—This odd state of mind, often called shamefacedness, or false shame, or *mauvaise honte*, appears to be one of the most efficient of all the causes of blushing. Shyness is, indeed, chiefly recognized by the face reddening, by the eyes being averted or cast down, and by awkward, nervous movements of the body. Many a woman blushes from this cause, a hundred, perhaps a thousand times, to once that she blushes from having done anything deserving blame, and of which she is truly ashamed. Shyness seems to depend on sensitiveness to the opinion, whether good or bad, of others, more especially with respect to external appearance. Strangers neither know nor care anything about our conduct or character, but they may, and often do, criticize our appearance: hence shy persons are particularly apt to be shy and to blush in the presence of strangers. The consciousness of anything peculiar, or even new, in the dress, or any slight blemish on the person, and more especially on the face—points which are likely to attract the attention of strangers—makes the shy intolerably shy. On the other hand, in those cases in which conduct and not personal appearance is concerned, we are much more apt to be shy in the presence of acquaintances, whose judgment we in some degree value, than in that of strangers. A physician told me that a young man, a wealthy duke, with whom he had travelled as medical attendant, blushed like a girl, when he paid him his fee; yet this young man probably would not have blushed and been shy, had he been paying a bill to a tradesman. Some persons, however, are so sensitive, that the mere act of speaking to almost any one is sufficient to rouse their self-consciousness, and a slight blush is the result.

Disapprobation or ridicule, from our sensitiveness on this head, causes shyness and blushing much more readily than does approbation; though the latter with some persons is highly efficient. The conceited are rarely shy; for they value themselves much too highly to expect depreciation. Why a proud man is often shy, as appears to be the case, is not so obvious, unless it be that, with all his self-reliance, he really thinks much about the opinion of others, although in a disdainful spirit. Persons who are exceedingly shy are rarely shy in the presence of those with whom they are quite familiar, and of whose good opinion and sympathy they are perfectly assured;—for instance, a girl in the presence of her mother. I neglected to inquire in my printed paper whether shyness can be detected in the different races of man; but a Hindoo gentleman assured Mr. Erskine that it is recognizable in his countrymen.

Shyness, as the derivation of the word indicates in several languages,[27] is closely related to fear; yet it is distinct from fear in the ordinary sense. A shy man no doubt dreads the notice of strangers, but can hardly be said to be afraid of them; he may be as bold as a hero in battle, and yet have no self-confidence about trifles in the presence of strangers. Almost every one is extremely nervous when first addressing a public assembly, and most men remain so throughout their lives; but this appears to depend on the consciousness of a

[27] H. Wedgwood, Dict. English Etymology, vol. iii. 1865, p. 184. So with the Latin word *verecundus*.

great coming exertion, with its associated effects on the system, rather than on shyness;[28] although a timid or shy man no doubt suffers on such occasions infinitely more than another. With very young children it is difficult to distinguish between fear and shyness; but this latter feeling with them has often seemed to me to partake of the character of the wildness of an untamed animal. Shyness comes on at a very early age. In one of my own children, when two years and three months old, I saw a trace of what certainly appeared to be shyness, directed towards myself after an absence from home of only a week. This was shown not by a blush, but by the eyes being for a few minutes slightly averted from me. I have noticed on other occasions that shyness or shamefacedness and real shame are exhibited in the eyes of young children before they have acquired the power of blushing.

As shyness apparently depends on self-attention, we can perceive how right are those who maintain that reprehending children for shyness, instead of doing them any good, does much harm, as it calls their attention still more closely to themselves. It has been well urged that "nothing hurts young people more than to be watched continually about their feelings, to have their countenances scrutinized, and the degrees of their sensibility measured by the surveying eye of the unmerciful spectator. Under the constraint of such examinations they can think of nothing but that they are looked at, and feel nothing but shame or apprehension."[29]

---

[28] Mr. Bain ('The Emotions and the Will,' p. 64) has discussed the "abashed" feelings experienced on these occasions, as well as the *stage-fright* of actors unused to the stage. Mr. Bain apparently attributes these feelings to simple apprehension or dread.

[29] 'Essays on Practical Education,' by Maria and R. L. Edgeworth, new edit. vol. ii. 1822, p. 38. Dr. Burgess (ibid. p. 187) insists strongly to the same effect.

Moral causes: guilt.—With respect to blushing from strictly moral causes, we meet with the same fundamental principle as before, namely, regard for the opinion of others. It is not the conscience which raises a blush, for a man may sincerely regret some slight fault committed in solitude, or he may suffer the deepest remorse for an undetected crime, but he will not blush. "I blush," says Dr. Burgess,[30] "in the presence of my accusers." It is not the sense of guilt, but the thought that others think or know us to be guilty which crimsons the face. A man may feel thoroughly ashamed at having told a small falsehood, without blushing; but if he even suspects that he is detected he will instantly blush, especially if detected by one whom he reveres.

On the other hand, a man may be convinced that God witnesses all his actions, and he may feel deeply conscious of some fault and pray for forgiveness; but this will not, as a lady who is a great blusher believes, ever excite a blush. The explanation of this difference between the knowledge by God and man of our actions lies, I presume, in man's disapprobation of immoral conduct being somewhat akin in nature to his depreciation of our personal appearance, so that through association both lead to similar results; whereas the disapprobation of God brings up no such association.

Many a person has blushed intensely when accused of some crime, though completely innocent of it. Even the thought, as the lady before referred to has observed to me, that others think that we have made an unkind or stupid remark, is amply sufficient to cause a blush, although we know all the time that we have been completely misunderstood. An action may be meritorious or of an indifferent nature, but a sensitive person, if he

---

[29] 'Essays on Practical Education,' by Maria and R. L. Edgeworth, new edit. vol. ii. 1822, p. 50.

suspects that others take a different view of it, will blush. For instance, a lady by herself may give money to a beggar without a trace of a blush, but if others are present, and she doubts whether they approve, or suspects that they think her influenced by display, she will blush. So it will be, if she offers to relieve the distress of a decayed gentlewoman, more particularly of one whom she had previously known under better circumstances, as she cannot then feel sure how her conduct will be viewed. But such cases as these blend into shyness.

Breaches of etiquette.—The rules of *etiquette* always refer to conduct in the presence of, or towards others. They have no necessary connection with the moral sense, and are often meaningless. Nevertheless as they depend on the fixed custom of our equals and superiors, whose opinion we highly regard, they are considered almost as binding as are the laws of honour to a gentleman. Consequently the breach of the laws of etiquette, that is, any impoliteness or *gaucherie*, any impropriety, or an inappropriate remark, though quite accidental, will cause the most intense blushing of which a man is capable. Even the recollection of such an act, after an interval of many years, will make the whole body to tingle. So strong, also, is the power of sympathy that a sensitive person, as a lady has assured me, will sometimes blush at a flagrant breach of etiquette by a perfect stranger, though the act may in no way concern her.

Modesty.—This is another powerful agent in exciting blushes; but the word modesty includes very different states of the mind. It implies humility, and we often judge of this by persons being greatly pleased and blushing at slight praise, or by being annoyed at praise which seems to them too high according to their own humble standard of themselves. Blushing here has the usual signification of regard for the opinion of others. But modesty frequently relates to acts of indelicacy; and indelicacy is an affair of etiquette, as we clearly see with the nations that go altogether or nearly naked. He who is modest, and blushes easily at acts of this nature, does so because they are breaches of a firmly and wisely established etiquette. This is indeed shown by the derivation of the word *modest* from *modus*, a measure or standard of behaviour. A blush due to this form of modesty is, moreover, apt to be intense, because it generally relates to the opposite sex; and we have seen how in all cases our liability to blush is thus increased. We apply the term 'modest,' as it would appear, to those who have an humble opinion of themselves, and to those who are extremely sensitive about an indelicate word or deed, simply because in both cases blushes are readily excited, for these two frames of mind have nothing else in common. Shyness also, from this same cause, is often mistaken for modesty in the sense of humility.

Some persons flush up, as I have observed and have been assured, at any sudden and disagreeable recollection. The commonest cause seems to be the sudden remembrance of not having done something for another person which had been promised. In this case it may be that the thought passes half unconsciously through the mind, "What will he think of me?" and then the flush would partake of the nature of a true blush. But whether such flushes are in most cases due to the capillary circulation being affected, is very doubtful; for we must remember that almost every strong emotion, such as anger or great joy, acts on the heart, and causes the face to redden.

The fact that blushes may be excited in absolute solitude seems opposed to the view here taken, namely that the habit originally arose from thinking about what others think of us. Several ladies, who are great blushers, are unanimous in regard to solitude; and some of them believe that they have blushed in the dark. From what Mr. Forbes has stated with respect to the Aymaras, and from my own sensations, I have no doubt that this latter statement is correct. Shakspeare, therefore, erred when he made Juliet, who was not even by herself, say to Romeo (act ii. sc. 2):—

> "Thou know'st the mask of night is on my face;
> Else would a maiden blush bepaint my cheek,
> For that which thou hast heard me speak to-night."

But when a blush is excited in solitude, the cause almost always relates to the thoughts of others about us—to acts done in their presence, or suspected by them; or again when we reflect what others would have thought of us had they known of the act. Nevertheless one or two of my informants believe that they have blushed from shame at acts in no way relating to others. If this be so, we must attribute the result to the force of inveterate habit and association, under a state of mind closely analogous to that which ordinarily excites a blush; nor need we feel surprise at this, as even sympathy with another person who commits a flagrant breach of etiquette is believed, as we have just seen, sometimes to cause a blush.

Finally, then, I conclude that blushing,—whether due to shyness—to shame for a real crime—to shame from a breach of the laws of etiquette—to modesty from humility—to modesty from an indelicacy—depends in all cases on the same principle; this principle being a sensitive regard for the opinion, more particularly for the depreciation of others, primarily in relation to our personal appearance, especially of our faces; and secondarily, through the force of association and habit, in relation to the opinion of others on our conduct.

Theory of Blushing.—We have now to consider, why should the thought that others are thinking about us affect our capillary circulation? Sir C. Bell insists[31] that blushing "is a provision for expression, as may be inferred from the colour extending only to the surface of the face, neck, and breast, the parts most exposed. It is not acquired; it is from the beginning." Dr. Burgess believes that it was designed by the Creator in "order that the soul might have sovereign power of displaying in the cheeks the various internal emotions of the moral feelings;" so as to serve as a check on ourselves, and as a sign to others, that we were violating rules which ought to be held sacred. Gratiolet merely remarks,—"Or, comme il est dans l'ordre de la nature que l'être social le plus intelligent soit aussi le plus intelligible, cette faculté de rougeur et de pâleur qui distingue l'homme, est un signe naturel de sa haute perfection."

The belief that blushing was *specially* designed by the Creator is opposed to the general theory of evolution, which is now so largely accepted; but it forms no part of my duty here to argue on the general question. Those who believe in design, will find it difficult to account for shyness being the most frequent and efficient of all the causes of blushing, as it makes the blusher to suffer and the beholder uncomfortable, without being of the least service to either of them. They will also find it difficult to account for negroes and other dark-coloured races blushing, in whom a change of colour in the skin is scarcely or not at all visible.

[31] Bell, 'Anatomy of Expression,' p. 95. Burgess, as quoted below, ibid. p. 49. Gratiolet, De la Phys. p. 94.

No doubt a slight blush adds to the beauty of a maiden's face; and the Circassian women who are capable of blushing, invariably fetch a higher price in the seraglio of the Sultan than less susceptible women.[32] But the firmest believer in the efficacy of sexual selection will hardly suppose that blushing was acquired as a sexual ornament. This view would also be opposed to what has just been said about the dark-coloured races blushing in an invisible manner.

The hypothesis which appears to me the most probable, though it may at first seem rash, is that attention closely directed to any part of the body tends to interfere with the ordinary and tonic contraction of the small arteries of that part. These vessels, in consequence, become at such times more or less relaxed, and are instantly filled with arterial blood. This tendency will have been much strengthened, if frequent attention has been paid during many generations to the same part, owing to nerve-force readily flowing along accustomed channels, and by the power of inheritance. Whenever we believe that others are depreciating or even considering our personal appearance, our attention is vividly directed to the outer and visible parts of our bodies; and of all such parts we are most sensitive about our faces, as no doubt has been the case during many past generations. Therefore, assuming for the moment that the capillary vessels can be acted on by close attention, those of the face will have become eminently susceptible. Through the force of association, the same effects will tend to follow whenever we think that others are considering or censuring our actions or character.

As the basis of this theory rests on mental attention having some power to influence the capillary circula-

[32] On the authority of Lady Mary Wortley Montague; see Burgess, ibid. p. 43.

tion, it will be necessary to give a considerable body of details, bearing more or less directly on this subject. Several observers,[33] who from their wide experience and knowledge are eminently capable of forming a sound judgment, are convinced that attention or consciousness (which latter term Sir H. Holland thinks the more explicit) concentrated on almost any part of the body produces some direct physical effect on it. This applies to the movements of the involuntary muscles, and of the voluntary muscles when acting involuntarily,—to the secretion of the glands,—to the activity of the senses and sensations,—and even to the nutrition of parts.

It is known that the involuntary movements of the heart are affected if close attention be paid to them. Gratiolet [34] gives the case of a man, who by continually watching and counting his own pulse, at last caused one beat out of every six to intermit. On the other hand, my father told me of a careful observer, who certainly had heart-disease and died from it, and who positively stated that his pulse was habitually irregular to an extreme degree; yet to his great disappointment it

[33] In England, Sir H. Holland was, I believe, the first to consider the influence of mental attention on various parts of the body, in his 'Medical Notes and Reflections,' 1839, p. 64. This essay, much enlarged, was reprinted by Sir H. Holland in his 'Chapters on Mental Physiology,' 1858, p. 79, from which work I always quote. At nearly the same time, as well as subsequently, Prof. Laycock discussed the same subject: see 'Edinburgh Medical and Surgical Journal,' 1839, July, pp. 17--22. Also his 'Treatise on the Nervous Diseases of Women,' 1840, p. 110; and 'Mind and Brain,' vol. ii. 1860, p. 327. Dr. Carpenter's views on mesmerism have a nearly similar bearing. The great physiologist Müller treated ('Elements of Physiology,' Eng. translat. vol. ii. pp. 937, 1085) of the influence of the attention on the senses. Sir J. Paget discusses the influence of the mind on the nutrition of parts, in his 'Lectures on Surgical Pathology,' 1853, vol. i. p. 39: I quote from the 3rd edit. revised by Prof. Turner, 1870, p. 28. See, also, Gratiolet, De la Phys. pp. 283--287.

[34] De la Phys. p. 283.

invariably became regular as soon as my father entered the room. Sir H. Holland remarks,[35] that "the effect upon the circulation of a part from the consciousness suddenly directed and fixed upon it, is often obvious and immediate." Professor Laycock, who has particularly attended to phenomena of this nature,[36] insists that "when the attention is directed to any portion of the body, innervation and circulation are excited locally, and the functional activity of that portion developed."

It is generally believed that the peristaltic movements of the intestines are influenced by attention being paid to them at fixed recurrent periods; and these movements depend on the contraction of unstriped and involuntary muscles. The abnormal action of the voluntary muscles in epilepsy, chorea, and hysteria is known to be influenced by the expectation of an attack, and by the sight of other patients similarly affected.[37] So it is with the involuntary acts of yawning and laughing.

Certain glands are much influenced by thinking of them, or of the conditions under which they have been habitually excited. This is familiar to every one in the increased flow of saliva, when the thought, for instance, of intensely acid fruit is kept before the mind.[38] It was shown in our sixth chapter, that an earnest and long-continued desire either to repress, or to increase, the action of the lacrymal glands is effectual. Some curious cases have been recorded in the case of women, of the power of the mind on the mammary glands; and still more remarkable ones in relation to the uterine functions.[39]

---

[35] 'Chapters on Mental Physiology,' 1858, p. 111.

[36] 'Mind and Brain,' vol. ii. 1860, p. 327.

[37] 'Chapters on Mental Physiology,' pp. 104--106.

[38] See Gratiolet on this subject, De la Phys. p. 287.

[39] Dr. J. Crichton Browne, from his observations on the insane, is convinced that attention directed for a prolonged

When we direct our whole attention to any one sense, its acuteness is increased;[40] and the continued habit of close attention, as with blind people to that of hearing, and with the blind and deaf to that of touch, appears to improve the sense in question permanently. There is, also, some reason to believe, judging from the capacities of different races of man, that the effects are inherited. Turning to ordinary sensations, it is well known that pain is increased by attending to it; and Sir B. Brodie goes so far as to believe that pain may be felt in any part of the body to which attention is closely drawn.[41] Sir H. Holland also remarks that we become not only conscious of the existence of a part subjected to concentrated attention, but we experience in it various odd sensations, as of weight, heat, cold, tingling, or itching.[42]

Lastly, some physiologists maintain that the mind

---

period on any part or organ may ultimately influence its capillary circulation and nutrition. He has given me some extraordinary cases; one of these, which cannot here be related in full, refers to a married woman fifty years of age, who laboured under the firm and long-continued delusion that she was pregnant. When the expected period arrived, she acted precisely as if she had been really delivered of a child, and seemed to suffer extreme pain, so that the perspiration broke out on her forehead. The result was that a state of things returned, continuing for three days, which had ceased during the six previous years. Mr. Braid gives, in his 'Magic, Hypnotism,' &c., 1852, p. 95, and in his other works analogous cases, as well as other facts showing the great influence of the will on the mammary glands, even on one breast alone.

[40] Dr. Maudsley has given ('The Physiology and Pathology of Mind,' 2nd edit. 1868, p. 105), on good authority, some curious statements with respect to the improvement of the sense of touch by practice and attention. It is remarkable that when this sense has thus been rendered more acute at any point of the body, for instance, in a finger, it is likewise improved at the corresponding point on the opposite side of the body.

[41] 'The Lancet,' 1838, pp. 39--40, as quoted by Prof. Laycock, 'Nervous Diseases of Women,' 1840, p. 110.

[42] 'Chapters on Mental Physiology,' 1858, pp. 91--93.

can influence the nutrition of parts. Sir J. Paget has given a curious instance of the power, not indeed of the mind, but of the nervous system, on the hair. A lady "who is subject to attacks of what is called nervous headache, always finds in the morning after such an one, that some patches of her hair are white, as if powdered with starch. The change is effected in a night, and in a few days after, the hairs gradually regain their dark brownish colour." [43]

We thus see that close attention certainly affects various parts and organs, which are not properly under the control of the will. By what means attention—perhaps the most wonderful of all the wondrous powers of the mind—is effected, is an extremely obscure subject. According to Müller,[44] the process by which the sensory cells of the brain are rendered, through the will, susceptible of receiving more intense and distinct impressions, is closely analogous to that by which the motor cells are excited to send nerve-force to the voluntary muscles. There are many points of analogy in the action of the sensory and motor nerve-cells; for instance, the familiar fact that close attention to any one sense causes fatigue, like the prolonged exertion of any one muscle.[45] When therefore we voluntarily concentrate our attention on any part of the body, the cells of the brain which receive impressions or sensations from that part are, it is probable, in some unknown manner stimulated into activity. This may account, without any local change in the part to which our attention is earnestly directed, for pain or odd sensations being there felt or increased.

---

[43] 'Lectures on Surgical Pathology,' 3rd edit. revised by Prof. Turner, 1870, pp. 28, 31.

[44] 'Elements of Physiology,' Eng. translat. vol. ii. p. 938.

[45] Prof. Laycock has discussed this point in a very interesting manner. See his 'Nervous Diseases of Women,' 1840, p. 110.

If, however, the part is furnished with muscles, we cannot feel sure, as Mr. Michael Foster has remarked to me, that some slight impulse may not be unconsciously sent to such muscles; and this would probably cause an obscure sensation in the part.

In a large number of cases, as with the salivary and lacrymal glands, intestinal canal, &c., the power of attention seems to rest, either chiefly, or as some physiologists think, exclusively, on the vaso-motor system being affected in such a manner that more blood is allowed to flow into the capillaries of the part in question. This increased action of the capillaries may in some cases be combined with the simultaneously increased activity of the sensorium.

The manner in which the mind affects the vaso-motor system may be conceived in the following manner. When we actually taste sour fruit, an impression is sent through the gustatory nerves to a certain part of the sensorium; this transmits nerve-force to the vaso-motor centre, which consequently allows the muscular coats of the small arteries that permeate the salivary glands to relax. Hence more blood flows into these glands, and they secrete a copious supply of saliva. Now it does not seem an improbable assumption, that, when we reflect intently on a sensation, the same part of the sensorium, or a closely connected part of it, is brought into a state of activity, in the same manner as when we actually perceive the sensation. If so, the same cells in the brain will be excited, though, perhaps, in a less degree, by vividly thinking about a sour taste, as by perceiving it; and they will transmit in the one case, as in the other, nerve-force to the vaso-motor centre with the same results.

To give another, and, in some respects, more appropriate illustration. If a man stands before a hot fire, his face reddens. This appears to be due, as Mr. Michael Foster informs me, in part to the local action of the heat, and in part to a reflex action from the vaso-motor centres.[46] In this latter case, the heat affects the nerves of the face; these transmit an impression to the sensory cells of the brain, which act on the vaso-motor centre, and this reacts on the small arteries of the face, relaxing them and allowing them to become filled with blood. Here, again, it seems not improbable that if we were repeatedly to concentrate with great earnestness our attention on the recollection of our heated faces, the same part of the sensorium which gives us the consciousness of actual heat would be in some slight degree stimulated, and would in consequence tend to transmit some nerve-force to the vaso-motor centres, so as to relax the capillaries of the face. Now as men during endless generations have had their attention often and earnestly directed to their personal appearance, and especially to their faces, any incipient tendency in the facial capillaries to be thus affected will have become in the course of time greatly strengthened through the principles just referred to, namely, nerve-force passing readily along accustomed channels, and inherited habit. Thus, as it appears to me, a plausible explanation is afforded of the leading phenomena connected with the act of blushing.

Recapitulation.—Men and women, and especially the young, have always valued, in a high degree, their personal appearance; and have likewise regarded the appearance of others. The face has been the chief object of attention, though, when man aboriginally went naked,

---

[46] See, also, Mr. Michael Foster, on the action of the vaso-motor system, in his interesting Lecture before the Royal Institution, as translated in the 'Revue des Cours Scientifiques,' Sept. 25, 1869, p. 683.

the whole surface of his body would have been attended to. Our self-attention is excited almost exclusively by the opinion of others, for no person living in absolute solitude would care about his appearance. Every one feels blame more acutely than praise. Now, whenever we know, or suppose, that others are depreciating our personal appearance, our attention is strongly drawn towards ourselves, more especially to our faces. The probable effect of this will be, as has just been explained, to excite into activity that part of the sensorium which receives the sensory nerves of the face; and this will react through the vaso-motor system on the facial capillaries. By frequent reiteration during numberless generations, the process will have become so habitual, in association with the belief that others are thinking of us, that even a suspicion of their depreciation suffices to relax the capillaries, without any conscious thought about our faces. With some sensitive persons it is enough even to notice their dress to produce the same effect. Through the force, also, of association and inheritance our capillaries are relaxed, whenever we know, or imagine, that any one is blaming, though in silence, our actions, thoughts, or character; and, again, when we are highly praised.

On this hypothesis we can understand how it is that the face blushes much more than any other part of the body, though the whole surface is somewhat affected, more especially with the races which still go nearly naked. It is not at all surprising that the dark-coloured races should blush, though no change of colour is visible in their skins. From the principle of inheritance it is not surprising that persons born blind should blush. We can understand why the young are much more affected than the old, and women more than men; and why the opposite sexes especially excite each other's blushes. It becomes obvious why personal remarks should be particularly liable to cause blushing, and why the most powerful of all the causes is shyness; for shyness relates to the presence and opinion of others, and the shy are always more or less self-conscious. With respect to real shame from moral delinquencies, we can perceive why it is not guilt, but the thought that others think us guilty, which raises a blush. A man reflecting on a crime committed in solitude, and stung by his conscience, does not blush; yet he will blush under the vivid recollection of a detected fault, or of one committed in the presence of others, the degree of blushing being closely related to the feeling of regard for those who have detected, witnessed, or suspected his fault. Breaches of conventional rules of conduct, if they are rigidly insisted on by our equals or superiors, often cause more intense blushes even than a detected crime; and an act which is really criminal, if not blamed by our equals, hardly raises a tinge of colour on our cheeks. Modesty from humility, or from an indelicacy, excites a vivid blush, as both relate to the judgment or fixed customs of others.

From the intimate sympathy which exists between the capillary circulation of the surface of the head and of the brain, whenever there is intense blushing, there will be some, and often great, confusion of mind. This is frequently accompanied by awkward movements, and sometimes by the involuntary twitching of certain muscles.

As blushing, according to this hypothesis, is an indirect result of attention, originally directed to our personal appearance, that is to the surface of the body, and more especially to the face, we can understand the meaning of the gestures which accompany blushing throughout the world. These consist in hiding the face, or turning it towards the ground, or to one side. The eyes are generally averted or are restless, for to look at the man who causes us to feel shame or shyness, immediately brings home in an intolerable manner the consciousness that his gaze is directed on us. Through the principle of associated habit, the same movements of the face and eyes are practised, and can, indeed, hardly be avoided, whenever we know or believe that others are blaming, or too strongly praising, our moral conduct.

CHAPTER XIV.

CONCLUDING REMARKS AND SUMMARY.

The three leading principles which have determined the chief movements of expression—Their inheritance—On the part which the will and intention have played in the acquirement of various expressions—The instinctive recognition of expression—The bearing of our subject on the specific unity of the races of man—On the successive acquirement of various expressions by the progenitors of man—The importance of expression—Conclusion.

I HAVE now described, to the best of my ability, the chief expressive actions in man, and in some few of the lower animals. I have also attempted to explain the origin or development of these actions through the three principles given in the first chapter. The first of these principles is, that movements which are serviceable in gratifying some desire, or in relieving some sensation, if often repeated, become so habitual that they are performed, whether or not of any service, whenever the same desire or sensation is felt, even in a very weak degree.

Our second principle is that of antithesis. The habit of voluntarily performing opposite movements under opposite impulses has become firmly established in us by the practice of our whole lives. Hence, if certain actions have been regularly performed, in accordance with our first principle, under a certain frame of mind, there will be a strong and involuntary tendency to the performance of directly opposite actions, whether or not these are of any use, under the excitement of an opposite frame of mind.

Our third principle is the direct action of the excited nervous system on the body, independently of the will, and independently, in large part, of habit. Experience shows that nerve-force is generated and set free whenever the cerebro-spinal system is excited. The direction which this nerve-force follows is necessarily determined by the lines of connection between the nerve-cells, with each other and with various parts of the body. But the direction is likewise much influenced by habit; inasmuch as nerve-force passes readily along accustomed channels.

The frantic and senseless actions of an enraged man may be attributed in part to the undirected flow of nerve-force, and in part to the effects of habit, for these actions often vaguely represent the act of striking. They thus pass into gestures included under our first principle; as when an indignant man unconsciously throws himself into a fitting attitude for attacking his opponent, though without any intention of making an actual attack. We see also the influence of habit in all the emotions and sensations which are called exciting; for they have assumed this character from having habitually led to energetic action; and action affects, in an indirect manner, the respiratory and circulatory system; and the latter reacts on the brain. Whenever these emotions or sensations are even slightly felt by us, though they may not at the time lead to any exertion, our whole system is nevertheless disturbed through the force of habit and association. Other emotions and sensations are called depressing, because they have not habitually led to energetic action, excepting just at first, as in the case of extreme pain, fear, and grief, and they have ultimately caused complete exhaustion; they are consequently expressed chiefly by negative signs and by prostration. Again, there are other emotions, such as that of affection, which do not commonly lead to action of any kind, and consequently are not exhibited by any strongly marked outward signs. Affection indeed, in as far as it is a pleasurable sensation, excites the ordinary signs of pleasure.

On the other hand, many of the effects due to the excitement of the nervous system seem to be quite independent of the flow of nerve-force along the channels which have been rendered habitual by former exertions of the will. Such effects, which often reveal the state of mind of the person thus affected, cannot at present be explained; for instance, the change of colour in the hair from extreme terror or grief,—the cold sweat and the trembling of the muscles from fear,—the modified secretions of the intestinal canal,—and the failure of certain glands to act.

Notwithstanding that much remains unintelligible in our present subject, so many expressive movements and actions can be explained to a certain extent through the above three principles, that we may hope hereafter to see all explained by these or by closely analogous principles.

Actions of all kinds, if regularly accompanying any state of the mind, are at once recognized as expressive. These may consist of movements of any part of the body, as the wagging of a dog's tail, the shrugging of a man's shoulders, the erection of the hair, the exudation of perspiration, the state of the capillary circulation, laboured breathing, and the use of the vocal or other sound-producing instruments. Even insects express anger, terror, jealousy, and love by their stridulation. With man the respiratory organs are of especial importance in expression, not only in a direct, but in a still higher degree in an indirect manner.

Few points are more interesting in our present subject than the extraordinarily complex chain of events which lead to certain expressive movements. Take, for instance, the oblique eyebrows of a man suffering from grief or anxiety. When infants scream loudly from hunger or pain, the circulation is affected, and the eyes tend to become gorged with blood: consequently the muscles surrounding the eyes are strongly contracted as a protection: this action, in the course of many generations, has become firmly fixed and inherited: but when, with advancing years and culture, the habit of screaming is partially repressed, the muscles round the eyes still tend to contract, whenever even slight distress is felt: of these muscles, the pyramidals of the nose are less under the control of the will than are the others, and their contraction can be checked only by that of the central fasciæ of the frontal muscle: these latter fasciæ draw up the inner ends of the eyebrows, and wrinkle the forehead in a peculiar manner, which we instantly recognize as the expression of grief or anxiety. Slight movements, such as these just described, or the scarcely perceptible drawing down of the corners of the mouth, are the last remnants or rudiments of strongly marked and intelligible movements. They are as full of significance to us in regard to expression, as are ordinary rudiments to the naturalist in the classification and genealogy of organic beings.

That the chief expressive actions, exhibited by man and by the lower animals, are now innate or inherited, —that is, have not been learnt by the individual,—is admitted by every one. So little has learning or imitation to do with several of them that they are from the earliest days and throughout life quite beyond our control; for instance, the relaxation of the arteries of the skin in blushing, and the increased action of the heart in anger. We may see children, only two or three years old, and even those born blind, blushing from shame; and the naked scalp of a very young infant reddens from passion. Infants scream from pain directly after birth, and all their features then assume the same form as during subsequent years. These facts alone suffice to show that many of our most important expressions have not been learnt; but it is remarkable that some, which are certainly innate, require practice in the individual, before they are performed in a full and perfect manner; for instance, weeping and laughing. The inheritance of most of our expressive actions explains the fact that those born blind display them, as I hear from the Rev. R. H. Blair, equally well with those gifted with eyesight. We can thus also understand the fact that the young and the old of widely different races, both with man and animals, express the same state of mind by the same movements.

We are so familiar with the fact of young and old animals displaying their feelings in the same manner, that we hardly perceive how remarkable it is that a young puppy should wag its tail when pleased, depress its ears and uncover its canine teeth when pretending to be savage, just like an old dog; or that a kitten should arch its little back and erect its hair when frightened and angry, like an old cat. When, however, we turn to less common gestures in ourselves, which we are accustomed to look at as artificial or conventional,—such as shrugging the shoulders, as a sign of impotence, or the raising the arms with open hands and extended fingers, as a sign of wonder,—we feel perhaps too much surprise at finding that they are innate. That these and some other gestures are inherited, we may infer from their being performed by very young children, by those born blind, and by the most widely distinct races of man. We should also bear in mind that new and highly peculiar tricks, in association with certain states of the mind, are known to have arisen in certain individuals, and to have been afterwards transmitted to their offspring, in some cases, for more than one generation.

Certain other gestures, which seem to us so natural that we might easily imagine that they were innate, apparently have been learnt like the words of a language. This seems to be the case with the joining of the uplifted hands, and the turning up of the eyes, in prayer. So it is with kissing as a mark of affection; but this is innate, in so far as it depends on the pleasure derived from contact with a beloved person. The evidence with respect to the inheritance of nodding and shaking the head, as signs of affirmation and negation, is doubtful; for they are not universal, yet seem too general to have been independently acquired by all the individuals of so many races.

We will now consider how far the will and consciousness have come into play in the development of the various movements of expression. As far as we can judge, only a few expressive movements, such as those just referred to, are learnt by each individual; that is, were consciously and voluntarily performed during the early years of life for some definite object, or in imitation of others, and then became habitual. The far greater number of the movements of expression, and all the more important ones, are, as we have seen, innate or inherited; and such cannot be said to depend on the will of the individual. Nevertheless, all those included under our first principle were at first voluntarily performed for a definite object,—namely, to escape some danger, to relieve some distress, or to gratify some desire. For instance, there can hardly be a doubt that the animals which fight with their teeth, have acquired the habit of drawing back their ears closely to their heads, when feeling savage, from their progenitors having voluntarily acted in this manner in order to protect their ears from being torn by their antagonists; for those animals which do not fight with their teeth do not thus express a savage state of mind. We may infer as highly probable that we ourselves have acquired the habit of contracting the muscles round the eyes, whilst crying gently, that is, without the utterance of any loud sound, from our progenitors, especially during infancy, having experienced, during the act of screaming, an uncomfortable sensation in their eyeballs. Again, some highly expressive movements result from the endeavour to check or prevent other expressive movements; thus the obliquity of the eyebrows and the drawing down of the corners of the mouth follow from the endeavour to prevent a screaming-fit from coming on, or to check it after it has come on. Here it is obvious that the consciousness and will must at first have come into play; not that we are conscious in these or in other such cases what muscles are brought into action, any more than when we perform the most ordinary voluntary movements.

With respect to the expressive movements due to the principle of antithesis, it is clear that the will has intervened, though in a remote and indirect manner. So again with the movements coming under our third principle; these, in as far as they are influenced by nerve-force readily passing along habitual channels, have been determined by former and repeated exertions of the will. The effects indirectly due to this latter agency are often combined in a complex manner, through the force of habit and association, with those directly resulting from the excitement of the cerebro-spinal system. This seems to be the case with the increased action of the heart under the influence of any strong emotion. When an animal erects its hair, assumes a threatening attitude, and utters fierce sounds, in order to terrify an enemy, we see a curious combination of movements which were originally voluntary with those that are involuntary. It is, however, possible that even strictly involuntary actions, such as the erection of the hair, may have been affected by the mysterious power of the will.

Some expressive movements may have arisen spontaneously, in association with certain states of the mind, like the tricks lately referred to, and afterwards been inherited. But I know of no evidence rendering this view probable.

The power of communication between the members of the same tribe by means of language has been of paramount importance in the development of man; and the force of language is much aided by the expressive movements of the face and body. We perceive this at once when we converse on an important subject with any person whose face is concealed. Nevertheless there are no grounds, as far as I can discover, for believing that any muscle has been developed or even modified exclusively for the sake of expression. The vocal and other sound-producing organs, by which various expressive noises are produced, seem to form a partial exception; but I have elsewhere attempted to show that these organs were first developed for sexual purposes, in order that one sex might call or charm the other. Nor can I discover grounds for believing that any inherited movement, which now serves as a means of expression, was at first voluntarily and consciously performed for this special purpose,—like some of the gestures and the finger-language used by the deaf and dumb. On the contrary, every true or inherited movement of expression seems to have had some natural and independent origin. But when once acquired, such movements may be voluntarily and consciously employed as a means of communication. Even infants, if carefully attended to, find out at a very early age that their screaming brings relief, and they soon voluntarily practise it. We may frequently see a person voluntarily raising his eyebrows to express surprise, or smiling to express pretended satisfaction and acquiescence. A man often wishes to make certain gestures conspicuous or demonstrative, and will raise his extended arms with widely opened fingers above his head, to show astonishment, or lift his shoulders to his ears, to show that he cannot or will not do something. The tendency to such movements will be strengthened or increased by their being thus voluntarily and repeatedly performed; and the effects may be inherited.

It is perhaps worth consideration whether movements at first used only by one or a few individuals to express a certain state of mind may not sometimes have spread to others, and ultimately have become universal, through the power of conscious and unconscious imitation. That there exists in man a strong tendency to imitation, independently of the conscious will, is certain. This is exhibited in the most extraordinary manner in certain brain diseases, especially at the commencement of inflammatory softening of the brain, and has been called the "echo sign." Patients thus affected imitate, without understanding, every absurd gesture which is made, and every word which is uttered near them, even in a foreign language.[1] In the case of animals, the jackal

[1] See the interesting facts given by Dr. Bateman on 'Aphasia,' 1870, p. 110.

and wolf have learnt under confinement to imitate the barking of the dog. How the barking of the dog, which serves to express various emotions and desires, and which is so remarkable from having been acquired since the animal was domesticated, and from being inherited in different degrees by different breeds, was first learnt, we do not know; but may we not suspect that imitation has had something to do with its acquisition, owing to dogs having long lived in strict association with so loquacious an animal as man?

In the course of the foregoing remarks and throughout this volume, I have often felt much difficulty about the proper application of the terms, will, consciousness, and intention. Actions, which were at first voluntary, soon became habitual, and at last hereditary, and may then be performed even in opposition to the will. Although they often reveal the state of the mind, this result was not at first either intended or expected. Even such words as that "certain movements serve as a means of expression" are apt to mislead, as they imply that this was their primary purpose or object. This, however, seems rarely or never to have been the case; the movements having been at first either of some direct use, or the indirect effect of the excited state of the sensorium. An infant may scream either intentionally or instinctively to show that it wants food; but it has no wish or intention to draw its features into the peculiar form which so plainly indicates misery; yet some of the most characteristic expressions exhibited by man are derived from the act of screaming, as has been explained.

Although most of our expressive actions are innate or instinctive, as is admitted by everyone, it is a different question whether we have any instinctive power of recognizing them. This has generally been assumed to be the case; but the assumption has been strongly controverted by M. Lemoine.[2] Monkeys soon learn to distinguish, not only the tones of voice of their masters, but the expression of their faces, as is asserted by a careful observer.[3] Dogs well know the difference between caressing and threatening gestures or tones; and they seem to recognize a compassionate tone. But as far as I can make out, after repeated trials, they do not understand any movement confined to the features, excepting a smile or laugh; and this they appear, at least in some cases, to recognize. This limited amount of knowledge has probably been gained, both by monkeys and dogs, through their associating harsh or kind treatment with our actions; and the knowledge certainly is not instinctive. Children, no doubt, would soon learn the movements of expression in their elders in the same manner as animals learn those of man. Moreover, when a child cries or laughs, he knows in a general manner what he is doing and what he feels; so that a very small exertion of reason would tell him what crying or laughing meant in others. But the question is, do our children acquire their knowledge of expression solely by experience through the power of association and reason?

As most of the movements of expression must have been gradually acquired, afterwards becoming instinctive, there seems to be some degree of *à priori* probability that their recognition would likewise have become instinctive. There is, at least, no greater difficulty in believing this than in admitting that, when a female quadruped first bears young, she knows the cry of distress of her offspring, or than in admitting that many animals instinctively recognize and fear their enemies; and of both these statements there can be no reason-

---

[2] 'La Physionomie et la Parole,' 1865, pp. 103, 118.

[3] Rengger, 'Naturgeschichte der Säugethiere von Paraguay,' 1830, s. 55.

able doubt. It is however extremely difficult to prove that our children instinctively recognize any expression. I attended to this point in my first-born infant, who could not have learnt anything by associating with other children, and I was convinced that he understood a smile and received pleasure from seeing one, answering it by another, at much too early an age to have learnt anything by experience. When this child was about four months old, I made in his presence many odd noises and strange grimaces, and tried to look savage; but the noises, if not too loud, as well as the grimaces, were all taken as good jokes; and I attributed this at the time to their being preceded or accompanied by smiles. When five months old, he seemed to understand a compassionate expression and tone of voice. When a few days over six months old, his nurse pretended to cry, and I saw that his face instantly assumed a melancholy expression, with the corners of the mouth strongly depressed; now this child could rarely have seen any other child crying, and never a grown-up person crying, and I should doubt whether at so early an age he could have reasoned on the subject. Therefore it seems to me that an innate feeling must have told him that the pretended crying of his nurse expressed grief; and this through the instinct of sympathy excited grief in him.

M. Lemoine argues that, if man possessed an innate knowledge of expression, authors and artists would not have found it so difficult, as is notoriously the case, to describe and depict the characteristic signs of each particular state of mind. But this does not seem to me a valid argument. We may actually behold the expression changing in an unmistakable manner in a man or animal, and yet be quite unable, as I know from experience, to analyse the nature of the change. In the two photographs given by Duchenne of the same old man (Plate

III. figs. 5 and 6), almost every one recognized that the one represented a true, and the other a false smile; but I have found it very difficult to decide in what the whole amount of difference consists. It has often struck me as a curious fact that so many shades of expression are instantly recognized without any conscious process of analysis on our part. No one, I believe, can clearly describe a sullen or sly expression; yet many observers are unanimous that these expressions can be recognized in the various races of man. Almost everyone to whom I showed Duchenne's photograph of the young man with oblique eyebrows (Plate II. fig. 2) at once declared that it expressed grief or some such feeling; yet probably not one of these persons, or one out of a thousand persons, could beforehand have told anything precise about the obliquity of the eyebrows with their inner ends puckered, or about the rectangular furrows on the forehead. So it is with many other expressions, of which I have had practical experience in the trouble requisite in instructing others what points to observe. If, then, great ignorance of details does not prevent our recognizing with certainty and promptitude various expressions, I do not see how this ignorance can be advanced as an argument that our knowledge, though vague and general, is not innate.

I have endeavoured to show in considerable detail that all the chief expressions exhibited by man are the same throughout the world. This fact is interesting, as it affords a new argument in favour of the several races being descended from a single parent-stock, which must have been almost completely human in structure, and to a large extent in mind, before the period at which the races diverged from each other. No doubt similar structures, adapted for the same purpose, have often been independently acquired through variation and natural selection by distinct species; but this view will not explain close similarity between distinct species in a multitude of unimportant details. Now if we bear in mind the numerous points of structure having no relation to expression, in which all the races of man closely agree, and then add to them the numerous points, some of the highest importance and many of the most trifling value, on which the movements of expression directly or indirectly depend, it seems to me improbable in the highest degree that so much similarity, or rather identity of structure, could have been acquired by independent means. Yet this must have been the case if the races of man are descended from several aboriginally distinct species. It is far more probable that the many points of close similarity in the various races are due to inheritance from a single parent-form, which had already assumed a human character.

It is a curious, though perhaps an idle speculation, how early in the long line of our progenitors the various expressive movements, now exhibited by man, were successively acquired. The following remarks will at least serve to recall some of the chief points discussed in this volume. We may confidently believe that laughter, as a sign of pleasure or enjoyment, was practised by our progenitors long before they deserved to be called human; for very many kinds of monkeys, when pleased, utter a reiterated sound, clearly analogous to our laughter, often accompanied by vibratory movements of their jaws or lips, with the corners of the mouth drawn backwards and upwards, by the wrinkling of the cheeks, and even by the brightening of the eyes.

We may likewise infer that fear was expressed from an extremely remote period, in almost the same manner as it now is by man; namely, by trembling, the erection of the hair, cold perspiration, pallor, widely opened eyes, the relaxation of most of the muscles, and by the whole body cowering downwards or held motionless.

Suffering, if great, will from the first have caused screams or groans to be uttered, the body to be contorted, and the teeth to be ground together. But our progenitors will not have exhibited those highly expressive movements of the features which accompany screaming and crying until their circulatory and respiratory organs, and the muscles surrounding the eyes, had acquired their present structure. The shedding of tears appears to have originated through reflex action from the spasmodic contraction of the eyelids, together perhaps with the eyeballs becoming gorged with blood during the act of screaming. Therefore weeping probably came on rather late in the line of our descent; and this conclusion agrees with the fact that our nearest allies, the anthropomorphous apes, do not weep. But we must here exercise some caution, for as certain monkeys, which are not closely related to man, weep, this habit might have been developed long ago in a sub-branch of the group from which man is derived. Our early progenitors, when suffering from grief or anxiety, would not have made their eyebrows oblique, or have drawn down the corners of their mouth, until they had acquired the habit of endeavouring to restrain their screams. The expression, therefore, of grief and anxiety is eminently human.

Rage will have been expressed at a very early period by threatening or frantic gestures, by the reddening of the skin, and by glaring eyes, but not by frowning. For the habit of frowning seems to have been acquired chiefly from the corrugators being the first muscles to contract round the eyes, whenever during infancy pain, anger, or distress is felt, and there consequently is a near approach to screaming; and partly from a frown serving as a shade in difficult and intent vision. It seems probable that this shading action would not have become habitual until man had assumed a completely upright position, for monkeys do not frown when exposed to a glaring light. Our early progenitors, when enraged, would probably have exposed their teeth more freely than does man, even when giving full vent to his rage, as with the insane. We may, also, feel almost certain that they would have protruded their lips, when sulky or disappointed, in a greater degree than is the case with our own children, or even with the children of existing savage races.

Our early progenitors, when indignant or moderately angry, would not have held their heads erect, opened their chests, squared their shoulders, and clenched their fists, until they had acquired the ordinary carriage and upright attitude of man, and had learnt to fight with their fists or clubs. Until this period had arrived the antithetical gesture of shrugging the shoulders, as a sign of impotence or of patience, would not have been developed. From the same reason astonishment would not then have been expressed by raising the arms with open hands and extended fingers. Nor, judging from the actions of monkeys, would astonishment have been exhibited by a widely opened mouth; but the eyes would have been opened and the eyebrows arched. Disgust would have been shown at a very early period by movements round the mouth, like those of vomiting,—that is, if the view which I have suggested respecting the source of the expression is correct, namely, that our progenitors had the power, and used it, of voluntarily and quickly rejecting any food from their stomachs which they disliked. But the more refined manner of showing contempt or disdain, by lowering the eyelids, or turning away the eyes and face, as if the despised person were not worth looking at, would not probably have been acquired until a much later period.

Of all expressions, blushing seems to be the most strictly human; yet it is common to all or nearly all the races of man, whether or not any change of colour is visible in their skin. The relaxation of the small arteries of the surface, on which blushing depends, seems to have primarily resulted from earnest attention directed to the appearance of our own persons, especially of our faces, aided by habit, inheritance, and the ready flow of nerve-force along accustomed channels; and afterwards to have been extended by the power of association to self-attention directed to moral conduct. It can hardly be doubted that many animals are capable of appreciating beautiful colours and even forms, as is shown by the pains which the individuals of one sex take in displaying their beauty before those of the opposite sex. But it does not seem possible that any animal, until its mental powers had been developed to an equal or nearly equal degree with those of man, would have closely considered and been sensitive about its own personal appearance. Therefore we may conclude that blushing originated at a very late period in the long line of our descent.

From the various facts just alluded to, and given in the course of this volume, it follows that, if the structure of our organs of respiration and circulation had differed in only a slight degree from the state in which they now exist, most of our expressions would have been wonderfully different. A very slight change in the course of the arteries and veins which run to the head, would probably have prevented the blood from accumulating in our eyeballs during violent expiration; for this occurs in extremely few quadrupeds. In this case we should not have displayed some of our most characteristic expressions. If man had breathed water by the aid of external branchiæ (though the idea is hardly conceivable), instead of air through his mouth and nostrils, his features would not have expressed his feelings much more efficiently than now do his hands or limbs. Rage and disgust, however, would still have been shown by movements about the lips and mouth, and the eyes would have become brighter or duller according to the state of the circulation. If our ears had remained movable, their movements would have been highly expressive, as is the case with all the animals which fight with their teeth; and we may infer that our early progenitors thus fought, as we still uncover the canine tooth on one side when we sneer at or defy any one, and we uncover all our teeth when furiously enraged.

The movements of expression in the face and body, whatever their origin may have been, are in themselves of much importance for our welfare. They serve as the first means of communication between the mother and her infant; she smiles approval, and thus encourages her child on the right path, or frowns disapproval. We readily perceive sympathy in others by their expression; our sufferings are thus mitigated and our pleasures increased; and mutual good feeling is thus strengthened. The movements of expression give vividness and energy to our spoken words. They reveal the thoughts and intentions of others more truly than do words, which may be falsified. Whatever amount of truth the so-called science of physiognomy may contain, appears to depend, as Haller long ago remarked,[4] on different persons bringing into frequent use different facial muscles, according

---

[4] Quoted by Moreau, in his edition of Lavater, 1820, tom. iv. p. 211.

to their dispositions; the development of these muscles being perhaps thus increased, and the lines or furrows on the face, due to their habitual contraction, being thus rendered deeper and more conspicuous. The free expression by outward signs of an emotion intensifies it. On the other hand, the repression, as far as this is possible, of all outward signs softens our emotions.[5] He who gives way to violent gestures will increase his rage; he who does not control the signs of fear will experience fear in a greater degree; and he who remains passive when overwhelmed with grief loses his best chance of recovering elasticity of mind. These results follow partly from the intimate relation which exists between almost all the emotions and their outward manifestations; and partly from the direct influence of exertion on the heart, and consequently on the brain. Even the simulation of an emotion tends to arouse it in our minds. Shakespeare, who from his wonderful knowledge of the human mind ought to be an excellent judge, says:—

> "Is it not monstrous that this player here,
> But in a fiction, in a dream of passion,
> Could force his soul so to his own conceit,
> That, from her working, all his visage wann'd;
> Tears in his eyes, distraction in 's aspect,
> A broken voice, and his whole function suiting
> With forms to his conceit? And all for nothing!"
> *Hamlet*, act ii. sc. 2.

We have seen that the study of the theory of expression confirms to a certain limited extent the conclusion that man is derived from some lower animal form, and supports the belief of the specific or subspecific unity of the several races; but as far as my judgment serves, such confirmation was hardly needed.

[5] Gratiolet ('De la Physionomie,' 1865, p. 66) insists on the truth of this conclusion.

We have also seen that expression in itself, or the language of the emotions, as it has sometimes been called, is certainly of importance for the welfare of mankind. To understand, as far as possible, the source or origin of the various expressions which may be hourly seen on the faces of the men around us, not to mention our domesticated animals, ought to possess much interest for us. From these several causes, we may conclude that the philosophy of our subject has well deserved the attention which it has already received from several excellent observers, and that it deserves still further attention, especially from any able physiologist.

INDEX.

THE END.